# Summation of Infinitely Small Quantities

# Summation of Infinitely Small Quantities

## I. P. Natanson

Dover Publications, Inc., Mineola, New York

*Bibliographical Note*

This Dover edition, first published in 2020, is an unabridged republication of the work originally published in 1963 by D. C. Heath and Company, Boston, as part of the "Topics in Mathematics" series. It was translated and adapted from the 1960 third Russian edition by Stephen Whelan and Coley Mills, Jr.

*Library of Congress Cataloging-in-Publication Data*

Names: Natanson, I. P. (Isidor Pavlovich), author.
Title: Summation of infinitely small quantities / I.P. Natanson.
Other titles: Summirovanie beskonechno malykh velichin. English
Description: Mineola, New York : Dover Publications, Inc., 2020. | This Dover edition, first published in 2020, is an unabridged republication of the work originally published in 1963 by D. C. Heath and Company, Boston, as part of the "Topics in Mathematics" series. It was translated and adapted from the 1960 third Russian edition by Stephen Whelan and Coley Mills, Jr. | Summary: "Translated and adapted from a popular Russian educational series, this concise book requires only some background in high school algebra and elementary trigonometry. It explores the fundamental concept of the integral calculus: the limit of the sum of an infinitely increasing number of infinitely decreasing quantities. Mastery of this concept enables the solution of geometry and physics problems and introduces the systematic study of higher mathematics"— Provided by publisher.
Identifiers: LCCN 2019054432 | ISBN 9780486843377 (trade paperback)
Subjects: LCSH: Calculus, Integral.
Classification: LCC QA308 .N313 2020 | DDC 515/.43—dc23
LC record available at https://lccn.loc.gov/2019054432

Manufactured in the United States by LSC Communications
84337801
www.doverpublications.com

2 4 6 8 10 9 7 5 3 1
2020

## PREFACE TO THE AMERICAN EDITION

IN ITS PRESENT FORM, integral calculus is a rather complex subject, since it is the result of an interplay of many very different ideas. Nevertheless, the fundamental concept upon which the integral calculus is built is quite simple and natural and was in essence known in antiquity. It is the concept of the "limit of the sum of an infinitely increasing number of infinitely decreasing quantities."

Mastering this concept is very useful, since it allows the solution of a number of important problems in geometry and physics, permits a deeper grasp of the idea of a limit, and serves as an excellent introduction to the systematic study of higher mathematics.

This booklet presents implications of this concept and their use in the solution of various concrete problems. For the first five chapters the reader should have a background of two years of high school algebra. A knowledge of elementary trigonometry is needed for an understanding of the last chapter.

# CONTENTS

# Summation of Infinitely Small Quantities

# 1. Some Algebraic Formulas

## 1. INTRODUCTION

In the presentation which follows, we shall need certain formulas, algebraic in nature, which are not always explained at school. These formulas give expressions for sums of the form

$$S_p = 1^p + 2^p + 3^p + \cdots + n^p,$$

where $p$ and $n$ denote positive whole numbers, and where the dots indicate that we keep adding the numbers $1^p$, $2^p$, $3^p$, etc., until we reach $n^p$. We require expressions for the sum $S_p$ only for small values of $p$:[1]

$$p = 1, 2, 3.$$

Let us derive these formulas.

## 2. THE SUM OF THE FIRST n NATURAL NUMBERS

Let us find, first of all, the sum

$$S_1 = 1 + 2 + 3 + \cdots + n.$$

This sum is the sum of $n$ terms of the arithmetic progression whose first term is $a_1 = 1$ and whose difference is $d = 1$; its value, therefore, can be determined with the help of the well-known algebraic formula

$$S_1 = \frac{n(n + 1)}{2}. \tag{1}$$

This formula can be derived as follows:

$$S_1 = 1 + 2 + 3 + \cdots + (n - 1) + n,$$
$$S_1 = n + (n - 1) + (n - 2) + \cdots + 2 + 1.$$

---

[1] The small $p$, 1, etc., in $S_p$, $S_1$, etc., are called *subscripts;* see section 6.

Adding these two equations term by term, we get

$$2S_1 = (n + 1) + (n + 1) + (n + 1) + \cdots + (n + 1) + (n + 1),$$

so that

$$2S_1 = n(n + 1),$$

from which formula (1) follows immediately.

We shall indicate another method for deriving formula (1), which, although a little more complicated than the method just employed, can be applied very well for finding any sum $S_p$ in the derivation of formulas $S_p$ (even when $p$ is greater than 1). Let us consider the equality

$$(n + 1)^2 = n^2 + 2n + 1$$

and in it successively replace $n$ by $n - 1$, then by $n - 2$, and so on until we reach 1. As a result we obtain a whole series of equalities

$$\left.\begin{aligned}
(n + 1)^2 &= n^2 &&+ 2n &&+ 1 \\
n^2 &= (n - 1)^2 &&+ 2(n - 1) &&+ 1 \\
(n - 1)^2 &= (n - 2)^2 &&+ 2(n - 2) &&+ 1 \\
&\cdots\cdots\cdots\cdots\cdots\cdots \\
2^2 &= 1^2 &&+ 2 \cdot 1 &&+ 1
\end{aligned}\right\} \qquad (2)$$

Let us add all these equalities. Notice that the column of terms on the left-hand side will be composed of almost the same terms as the column of *first* terms on the right-hand side. The only differences between the two columns are these: the term $1^2$, which stands last in the column on the right, does not appear in the left column; and the term $(n + 1)^2$, which stands first in the left column is absent from the right side.

Having made this observation, we see that, cancelling the same terms of both columns, we get

$$(n + 1)^2 = 1^2 + \{2n + 2(n - 1) + \cdots + 2 \cdot 1\}$$
$$+ \{1 + 1 + \cdots + 1\}.$$

The number of terms in the *second* pair of braces is equal to the number of equalities in system (2). Since there are $n$ equalities, the sum indicated in this pair of braces is simply $n$. Observe, moreover, that if we take the common factor 2 out of the expression enclosed by the *first* pair of braces, the expression remaining in the braces is precisely the sum $S_1$. If we further replace $1^2$ by 1, we get

$$(n + 1)^2 = 1 + 2S_1 + n.$$

Hence,

$$2S_1 = (n + 1)^2 - (n + 1) = (n + 1)[(n + 1) - 1] = n(n + 1),$$

and, finally,

$$S_1 = \frac{n(n + 1)}{2},$$

so that we get formula (1) again.

## 3. THE SUM OF THE SQUARES

Let us now adopt a similar method for computing the sum of the *squares* of the first $n$ natural numbers, that is, the sum

$$S_2 = 1^2 + 2^2 + 3^2 + \cdots + n^2.$$

To this end, we successively replace $n$ by $n - 1$, by $n - 2$, and so on, in the equality

$$(n + 1)^3 = n^3 + 3n^2 + 3n + 1,$$

until, in the final stage, we replace $n$ by 1. We obtain the following system of equalities:

$$\left.\begin{aligned}
(n + 1)^3 &= n^3 &&+ 3n^2 &&+ 3n &&+ 1 \\
n^3 &= (n - 1)^3 &&+ 3(n - 1)^2 &&+ 3(n - 1) &&+ 1 \\
(n - 1)^3 &= (n - 2)^3 &&+ 3(n - 2)^2 &&+ 3(n - 2) &&+ 1 \\
&\,\cdot\ \cdot\ \cdot\ \cdot\ \cdot\ \cdot\ \cdot\ \cdot\ \cdot\ \cdot\ \cdot\ \cdot\ \cdot\ \cdot\ \cdot \\
2^3 &= 1^3 &&+ 3 \cdot 1^2 &&+ 3 \cdot 1 &&+ 1.
\end{aligned}\right\} \quad (3)$$

Let us add all these equalities. As in the previous case, we can simplify considerably as follows: From the column of terms on the left-hand side we can cancel all the terms except the first one, that is, except $(n + 1)^3$; and, from the column of *first* terms on the right-hand side we can cancel all the terms except the last one, that is, except $1^3$.

Furthermore, if from the column of *second* terms on the right-hand side we take out the common factor 3, then, clearly, we are left with precisely the desired sum $S_2$. In exactly the same way the column of *third* terms of the right-hand side gives three times the sum $S_1$, which we have found above. Observing further that the number of equalities in (3) is $n$, we get

$$(n + 1)^3 = 1^3 + 3S_2 + 3S_1 + n.$$

Now let us replace $1^3$ by 1, and $S_1$ by expression (1) to get

$$(n + 1)^3 = 1 + 3S_2 + 3\,\frac{n(n + 1)}{2} + n.$$

Thus,

$$3S_2 = (n + 1)^3 - \frac{3}{2}n(n + 1) - (n + 1),$$

or

$$3S_2 = (n + 1)\left[(n + 1)^2 - \frac{3}{2}n - 1\right] = n(n + 1)\left(n + \frac{1}{2}\right).$$

Therefore,

$$3S_2 = \frac{n(n + 1)(2n + 1)}{2}.$$

Finally, we have

$$S_2 = \frac{n(n + 1)(2n + 1)}{6}. \tag{4}$$

## 4. THE SUM OF THE CUBES

In exactly the same way, proceeding from the equality

$$(n + 1)^4 = n^4 + 4n^3 + 6n^2 + 4n + 1,$$

we arrive at the system of equalities:

$$\left.\begin{aligned}
(n + 1)^4 &= n^4 && + 4n^3 && + 6n^2 && + 4n && + 1, \\
n^4 &= (n - 1)^4 && + 4(n - 1)^3 && + 6(n - 1)^2 && + 4(n - 1) && + 1, \\
&\;\cdot\;\cdot\;\cdot\;\cdot\;\cdot\;\cdot\;\cdot\;\cdot\;\cdot\;\cdot\;\cdot\;\cdot\;\cdot\;\cdot\;\cdot\;\cdot\;\cdot\;\cdot\;\cdot\;\cdot\;\cdot \\
2^4 &= 1^4 && + 4 \cdot 1^3 && + 6 \cdot 1^2 && + 4 \cdot 1 && + 1.
\end{aligned}\right\}$$

After addition and simplification we get

$$(n + 1)^4 = 1 + 4S_3 + 6S_2 + 4S_1 + n.$$

Substituting for the sums $S_1$ and $S_2$ the expressions (1) and (4), and carrying out all the computations, we eventually obtain the following expression for the sum $S_3$:[1]

$$S_3 = \frac{n^2(n+1)^2}{4}. \tag{5}$$

In a similar way one can find the sums $S_4$, $S_5$, etc.

## 5. THE RELATIONSHIP BETWEEN $S_1$ AND $S_3$

Although it has no direct bearing on the theme of this booklet, we should like to mention one interesting relationship between formulas (1) and (5); namely, upon comparison of these formulas, we see that

$$S_3 = S_1{}^2$$

or, more explicitly,

$$1^3 + 2^3 + \cdots + n^3 = (1 + 2 + \cdots + n)^2. \tag{6}$$

For example,

$$1^3 + 2^3 = 9 \text{ and } (1 + 2)^2 = 9,$$

$$1^3 + 2^3 + 3^3 = 36 \text{ and } (1 + 2 + 3)^2 = 36,$$

and

$$1^3 + 2^3 + 3^3 + 4^3 = 100 \text{ and } (1 + 2 + 3 + 4)^2 = 100.$$

Equality (6) seems all the more interesting since it is easy to show that there is by no means a general equality

$$a^3 + b^3 + \cdots + k^3 = (a + b + \cdots + k)^2$$

that is valid for arbitrary values of $a, b, \ldots, k$. For example,

$$2^3 + 4^3 = 72, \qquad (2 + 4)^2 = 36,$$

but

$$72 \neq 36.$$

---

[1] We suggest that the reader convince himself of the correctness of (5) by performing these calculations on a sheet of scratch paper.

## 6. THE $\Sigma$ NOTATION

Formulas (1), (4), and (5) can be written in another form if we make use of the $\Sigma$ notation—a notation widely used in mathematics. For example, if we have a series of terms (the number of terms being, say, $n$), it is often convenient to denote these terms by means of a single letter accompanied by a numerical "suffix" (called a *subscript*) written to the lower right of the letter. Thus, a series of $n$ terms might be written as follows:

$$a_1, a_2, \ldots, a_n.$$

If we are interested in the sum of the terms

$$a_1, a_2, \ldots, a_n,$$

this sum can be denoted by

$$\sum_{k=1}^{n} a_k. \tag{7}$$

Thus,

$$\sum_{k=1}^{n} a_k = a_1 + a_2 + a_3 + \cdots + a_n.$$

The symbol $a_k$ in (7) is used to denote the general (or $k$th) term of the series. The numbers which appear above and below the summation sign $\Sigma$ tell us that the subscripts attached to the letter $a$ run through all of the positive whole numbers from 1 to $n$. Hence, one may read the symbol

$$\sum_{k=1}^{n} a_k$$

as "the sum of $a_k$ as $k$ runs from 1 to $n$." The symbol $\Sigma$ is the capital Greek letter *sigma*.

In order to illustrate the use of the notation, let us suppose that the number of terms in a certain series is 5. The terms of the series may be denoted by

$$a_1, a_2, a_3, a_4, a_5.$$

In this case $n = 5$ and the sum of all the terms is

$$\sum_{k=1}^{5} a_k = a_1 + a_2 + a_3 + a_4 + a_5.$$

Using the $\Sigma$ notation, the sums $S_1$, $S_2$, and $S_3$ can be expressed as follows:

$$S_1 = \sum_{k=1}^{n} k, \quad S_2 = \sum_{k=1}^{n} k^2, \quad S_3 = \sum_{k=1}^{n} k^3,$$

and formulas (1), (4), and (5) take the forms[1]

$$\sum_{k=1}^{n} k = \frac{n(n+1)}{2}, \tag{8}$$

$$\sum_{k=1}^{n} k^2 = \frac{n(n+1)(2n+1)}{6}, \tag{9}$$

$$\sum_{k=1}^{n} k^3 = \frac{n^2(n+1)^2}{4}. \tag{10}$$

## 7. SOME PROPERTIES OF THE $\Sigma$ NOTATION

Let us note some additional properties of the $\Sigma$ notation.

(1) If each of the terms is itself the sum of two terms, then their sum breaks up into two sums. In other words:

$$\sum_{k=1}^{n} (a_k + b_k) = \sum_{k=1}^{n} a_k + \sum_{k=1}^{n} b_k. \tag{11}$$

To prove equality (11) it is sufficient to write the left-hand side in expanded form as

$$(a_1 + b_1) + (a_2 + b_2) + \cdots + (a_n + b_n);$$

but clearly, this can be written as

$$(a_1 + a_2 + \cdots + a_n) + (b_1 + b_2 + \cdots + b_n),$$

and this is precisely the right-hand side of equality (11).

---

[1] We assume that the reader of this booklet will study it with pencil in hand. We recommend that he copy formulas (8), (9), and (10) on a separate sheet of paper and have them before him as he proceeds.

(2) If all the terms of a sum have a common factor, then this factor can be taken outside the summation sign:

$$\sum_{k=1}^{n} ca_k = c \sum_{k=1}^{n} a_k. \tag{12}$$

The proof is left to the reader.

(3) If all the terms $a_k$ are equal to *one and the same number a,* then the sum is equal to this same number multiplied by the number of terms:

$$\text{If } a_1 = a_2 = \cdots = a_n = a, \text{ then } \sum_{k=1}^{n} a_k = \sum_{k=1}^{n} a = na. \tag{13}$$

This property also can be easily proved by the reader. In view of the extreme simplicity of the properties that we have indicated for the $\Sigma$ notation, we shall henceforth make use of them without any further comment.

# 2. Determination of the Pressure of a Liquid on a Vertical Wall

## 8. INTRODUCTION

Assume that we have a rectangular tank filled with water; its dimensions are shown in Fig. 1. We pose the problem of finding the pressure[1] $P$ of the water on the front wall of the tank.

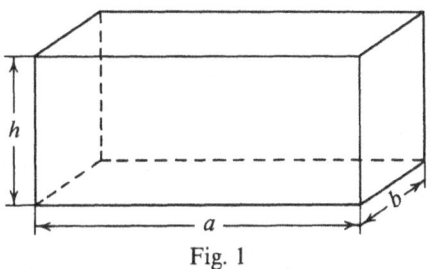

Fig. 1

In order to solve this problem, it is necessary to recall some laws of hydrostatics.

## 9. A LAW OF HYDROSTATICS

If we think of a *horizontal* plane surface under the water, then the pressure of the water on it is equal to the weight of the column of water resting on this plane surface, that is, of a column of water having this surface for its base and for its altitude the depth to which the surface is submerged. Since we are talking about water, the specific gravity[2] of which is equal to 1, the weight of the column concerned is numerically equal to its volume,[3] that is, the area of

---

[1] Both here and later, in speaking of "pressure," we have in mind the total force with which the water acts on the wall, and not the force per unit area (that is, not the specific pressure).

[2] The specific gravity of any substance is the ratio of the weight of a given volume of the substance to the weight of an equal volume of water.

[3] This is true if units in the metric system, such as cubic centimeters and grams, are used.

the plane surface multiplied by the depth to which it is submerged. This product, therefore, gives the amount of pressure on the horizontal surface.

If the submerged surface is not horizontal, then various points of it are situated at different depths and we cannot speak of the depth of submersion of the surface as a whole. But if this surface is *very small,* then it is possible to make *approximate* calculations by assuming that all the submerged points are at the same depth and by calling this depth the depth of submersion of the surface.

We shall assume that we are given just such a very small plane surface submerged in water, and we wish to determine the pressure on it. To do this, suppose we turn the surface about one of its points until the surface assumes a horizontal position. The pressure at any given point within the liquid is transmitted equally in all directions, and the dimensions of the surface are very small, and thus a slight change in the position of a point only slightly alters the pressure at that point, and the turning to which we have referred only slightly alters the pressure on the surface as a whole. Now that the surface is in a *horizontal* position, we can apply to it the rule indicated above for determining the pressure on the surface. Since turning the surface does not alter its area, and since its depth of submersion is not altered appreciably, we can make the following assertion: *If a very small plane surface is submerged in water, the pressure on this surface is equal to the area of the surface multiplied by its depth of submersion.*

This rule is not completely accurate, but only approximate. The smaller the area of the surface considered, the smaller the error of approximation.

## 10. THE PRESSURE ON THE WALL OF A RECTANGULAR TANK

Having established the above rule, let us now return to the problem posed in section 8. The front wall of the tank is not small and thus the rule we have established is not directly applicable to it. In order to apply the rule, let us proceed in the following manner.

Take a very large number $n$ and divide the wall into $n$ identical horizontal strips (Fig. 2) each of width $\frac{1}{n}h$. Now consider one of these strips, for example, the $k$th (from the top). It is very narrow and we can assume that all the points on it lie at approximately the same depth. Hence the pressure on the strip can be found with the help of the law stated in section 9.

The area of the strip is equal to the product of its length $a$ and its width $\frac{1}{n}h$; that is, the area is $\frac{1}{n}ah$. In order to calculate the pressure on this strip, it is necessary to multiply this area by the depth

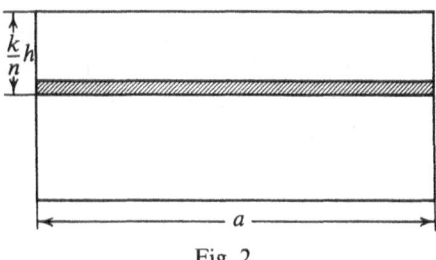

Fig. 2

of submersion of the strip. The depth of submersion of the $k$th strip is $\frac{k}{n}h$.[1] If we consider the derivation in section 9 of the pressure law, we see that the only essential assumption was that all points of the plane surface lie (at least approximately) at the same depth below the surface of the water. Consequently, it is quite legitimate to apply the law to the narrow horizontal strip in spite of the fact that, owing to its length, it is not "small." Therefore, the pressure $P_k$ on the $k$th strip is

$$P_k = \frac{ah^2}{n^2} k.$$

To determine the pressure $P$ on the whole wall, we add the pressures on the individual strips, and get

$$P = \sum_{k=1}^{n} \frac{ah^2}{n^2} k \quad \text{or} \quad P = \frac{ah^2}{n^2} \sum_{k=1}^{n} k.$$

[1] The quantity $\frac{k}{n}h$ is the depth of the *lower edge* of the $k$th strip, but since we have agreed to disregard the difference in depth of the various points of the strip, we take this quantity to be the depth of submersion of the entire strip. Later on we shall frequently have to deal with similar situations.

11

Using formula (8), we can express the pressure $P$ as

$$P = \frac{ah^2}{n^2} \cdot \frac{n(n+1)}{2}, \quad \text{or as} \quad P = \frac{ah^2}{2}\left(1 + \frac{1}{n}\right),$$

from which, finally, we get

$$P = \frac{ah^2}{2} + \frac{ah^2}{2} \cdot \frac{1}{n}. \tag{14}$$

However, the expression which we have found for the pressure is only approximate. For, in fact, although the strips are very narrow, different points, even in a given strip, lie at different depths.

To indicate that two numbers $A$ and $B$ are approximately equal, mathematicians often write

$$A \approx B.$$

Thus, "$A \approx B$" is to be read "$A$ is approximately equal to $B$." For example, it is correct to write $\sqrt{2} \approx 1.414$, but it is *not* correct to write $\sqrt{2} = 1.414$. This notation may be used to rewrite equation (14) as

$$P \approx \frac{ah^2}{2} + \frac{ah^2}{2} \cdot \frac{1}{n}. \tag{14'}$$

It is clear that (14') will become more accurate as we take smaller and smaller strips, that is, as we let $n$ increase. Thus, the correct value for the pressure is the limit,[1] which is the quantity that

$$\frac{ah^2}{2} + \frac{ah^2}{2} \cdot \frac{1}{n}$$

approaches as $n$ increases indefinitely. It is clear that as $n$ increases, the quantities $\frac{1}{n}$ and $\frac{ah^2}{2} \cdot \frac{1}{n}$ become smaller and smaller, tending to zero. Hence, the *limit* of the quantity $\frac{ah^2}{2} + \frac{ah^2}{2} \cdot \frac{1}{n}$ is the first term $\frac{ah^2}{2}$. This gives us a completely accurate expression for the pressure:

$$P = \frac{ah^2}{2}.$$

Thus the problem is solved.

[1] The limit of a variable quantity $x_n$ is a constant number $l$, such that the absolute value of the difference $x_n - l$, for all sufficiently large values of $n$, is less than any positive number whatever which we have chosen beforehand.

## 11. THE PRESSURE ON A TRIANGULAR PLATE, APEX DOWN

Let us now pose another problem of the same kind. Let us try to determine the pressure on a triangular plate vertically immersed in water in such a way that the base of the triangle is at the level of the surface of the liquid (Fig. 3).

In order to solve this problem, making use of the method presented in the preceding paragraphs, we again divide the plate into $n$ very narrow horizontal strips, each of width $\frac{1}{n}h$. We shall determine the total pressure on the triangular plate by calculating the sum of the pressures on the individual strips.

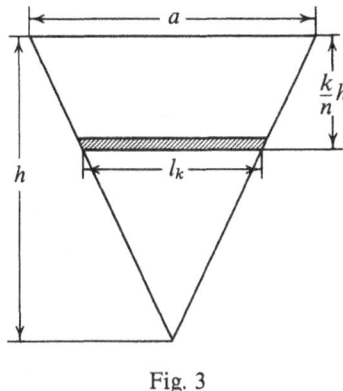

Fig. 3

Let us take a particular strip, the $k$th one from the top, and calculate the pressure exerted upon it. Disregarding the width of the strip, we can assume that all the points on it are at the same depth, $\frac{k}{n}h$, beneath the surface. The pressure $P_k$ on this strip is obtained by multiplying the depth by the area of the strip. This area is the area of a trapezoid. However, so long as the strip is narrow, we can simplify the calculation of the area by assuming that the shape is *rectangular*. Some error does indeed result from this, but, as in the previous example, it will turn out that this error is of no consequence. Here we meet an idea of a very general nature which is constantly employed for the solution of a great variety of problems: To calculate some quantity (such as the pressure on a submerged surface), find a simple way to perform the calculation, neglecting for the sake of simplicity small portions of the quantity, making sure, however, that the neglected portion is insignificant compared with what has been taken into account. Using the theory of limits, this principle can be expressed in a more accurate and rigorous form; however, this will not be done at this point, since the meaning of these remarks will be clarified by subsequent examples.

Taking the $k$th strip as a rectangle, we find its area by calculating the product of its length and its width. The width is, clearly, $\frac{1}{n} h$.

The length $l_k$ (the subscript $k$ shows that we are speaking of the $k$th strip) is found (see Fig. 3) from similar triangles by the proportion

$$l_k : a = \left( h - \frac{k}{n} h \right) : h,$$

whence

$$l_k = \left( 1 - \frac{k}{n} \right) a.$$

Thus, the area of the strip is

$$\left( 1 - \frac{k}{n} \right) a \cdot \frac{1}{n} h,$$

so that the pressure on it is

$$P_k \approx \frac{ah^2}{n^2} \left( 1 - \frac{k}{n} \right) k.$$

The total pressure on the triangular plate is found by summing the pressures on the individual strips:

$$P \approx \sum_{k=1}^{n} \frac{ah^2}{n^2} \left( 1 - \frac{k}{n} \right) k,$$

or

$$P \approx \frac{ah^2}{n^2} \sum_{k=1}^{n} k - \frac{ah^2}{n^3} \sum_{k=1}^{n} k^2.$$

Using formulas (8) and (9), we can rewrite the last approximation as

$$P \approx \frac{ah^2}{n^2} \cdot \frac{n(n+1)}{2} - \frac{ah^2}{n^3} \cdot \frac{n(n+1)(2n+1)}{6},$$

or

$$P \approx \frac{ah^2}{2} \left( 1 + \frac{1}{n} \right) - \frac{ah^2}{6} \left( 1 + \frac{1}{n} \right) \left( 2 + \frac{1}{n} \right).$$

This expression for the pressure is only approximate. The more we increase the number $n$, the more accurate our approximation becomes. This means that to find the exact value of $P$, it is necessary to increase $n$ indefinitely and to find the *limit* of the right-hand side. Since an increase in the number $n$ tends to make the fraction $\frac{1}{n}$ approach zero, the factors $\left(1 + \frac{1}{n}\right)$ and $\left(2 + \frac{1}{n}\right)$ tend respectively to approach 1 and 2; thus (on the basis of theorems about the limit of a product and of a difference) the entire expression has $\frac{ah^2}{2} - \frac{ah^2}{6} \cdot 2$ as its limit. Thus,

$$P = \frac{ah^2}{2} - \frac{ah^2}{3}$$

and, finally,

$$P = \frac{ah^2}{6}.$$

This is the *exact* value of the pressure.

## 12. THE PRESSURE ON A TRIANGULAR PLATE, APEX UP

Let us find the pressure on a vertical plate of the same shape, but immersed in water in such a way that its apex is at the surface of the water, and its base lies parallel to the surface (Fig. 4).

Dividing the plate into horizontal strips of width $\frac{1}{n}h$ and taking each strip to be a rectangle, we find the length of the $k$th strip, using similar triangles:

$$l_k : a = \frac{k}{n}h : h,$$

from which we get

$$l_k = \frac{k}{n}a.$$

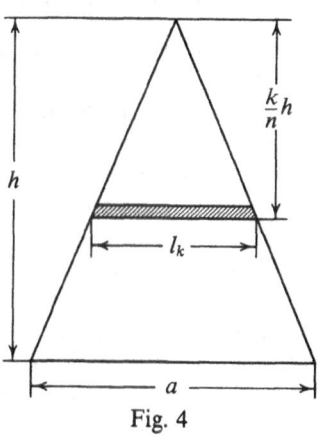

We see now that the area of the strip is equal to $\frac{k}{n^2}ah$ and, since the depth to which it is immersed is $\frac{k}{n}h$, the pressure on it is

$$P_k \approx \frac{k^2}{n^3}ah^2.$$

Fig. 4

15

The total pressure on the plate is obtained by summing all these pressures:

$$P \approx \sum_{k=1}^{n} \frac{k^2}{n^3} ah^2 = \frac{ah^2}{n^3} \sum_{k=1}^{n} k^2.$$

With the help of formula (9) we write $P$ as

$$P \approx \frac{ah^2}{n^3} \cdot \frac{n(n+1)(2n+1)}{6},$$

or as

$$P \approx \frac{ah^2}{6} \left(1 + \frac{1}{n}\right)\left(2 + \frac{1}{n}\right).$$

The exact expression is found from this by passing to the limit as $n$ increases indefinitely; to find this limit, it is necessary to repeat the argument outlined at the end of section 11. Without going into detail, we merely remark that we find the limit we are seeking by disregarding the term $\frac{1}{n}$ in the parentheses, obtaining

$$P = \frac{ah^2}{3}.$$

## 13. THE PRESSURE ON A SEMICIRCULAR PLATE

In the examples previously analyzed it is clear that the same idea has been followed throughout. It consists of dividing the desired pressure $P$ into the terms $P_k$. The calculation of one term is effected by a simple method (that is, by disregarding the difference in depth of the different points on one strip and taking the strip to be a rectangle), which enables us to find $P_k$ easily. Then all these pressures are summed and the limit of the resulting sum is determined by increasing $n$ indefinitely. To find the limit of the sum, we make use of formulas (8) and (9) of section 6. However, it would be erroneous to think that the solution of problems by the method indicated always leads to the simple sums of Chapter 1; on the contrary, we quite often arrive at far more complicated sums. We shall illustrate this by means of an example: Let us try to determine the pressure on a semicircular plate (Fig. 5) vertically immersed in water in such a way that the surface of the liquid coincides with the diameter of the semicircle.

Applying the principles already discussed, let us divide the plate into horizontal strips of width $\frac{1}{n} R$, where $R$ is the radius of the semi-

circle. Once again we take each strip to be a rectangle. Its length can be found by using the Pythagorean theorem:

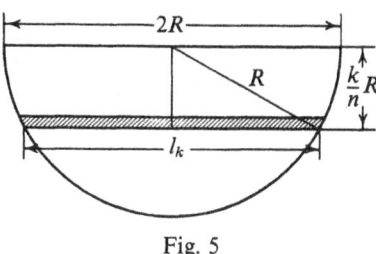

Fig. 5

$$l_k = 2 \sqrt{R^2 - \frac{k^2}{n^2} R^2}$$

$$= \frac{2R}{n} \sqrt{n^2 - k^2}.$$

In this case the area of the strip is

$$\frac{2R^2}{n^2} \sqrt{n^2 - k^2},$$

and the pressure is

$$P_k \approx \frac{2R^3}{n^3} k \sqrt{n^2 - k^2}.$$

The approximate expression for the total pressure is the sum

$$P \approx \sum_{k=1}^{n} \frac{2R^3}{n^3} k \sqrt{n^2 - k^2},$$

or

$$P \approx \frac{2R^3}{n^3} \sum_{k=1}^{n} k \sqrt{n^2 - k^2},$$

while the exact value of $P$ is the *limit* of this sum as $n$ increases indefinitely. Let us note that, strictly speaking, we are not interested in the sum itself but rather its limit. Thus,

$$P = 2R^3 \lim \left[ \frac{1}{n^3} \sum_{k=1}^{n} k \sqrt{n^2 - k^2} \right], \tag{15}$$

where the symbol "lim" is an abbreviation for "limit."

17

Consequently, the whole problem would be solved if we were able to find the limit

$$\lim \left[ \frac{1}{n^3} \sum_{k=1}^{n} k \sqrt{n^2 - k^2} \right]. \tag{16}$$

However, we cannot, at present, find this limit and hence cannot solve the given problem. Later on, in section 23, we present a technique for calculating the limit (16), and thereby solving our problem.

# 3. Calculation of the Work Done in Pumping Liquid from a Container

## 14. PUMPING WATER OUT OF A CYLINDRICAL BOILER

In this chapter we shall consider a type of problem which is related to a very different domain of physics but whose solution is arrived at by means of the same method of division into an infinitely growing number of infinitely decreasing, or, as we say, *infinitely small,* terms.

As a typical example let us consider the following problem. Suppose that a cylindrical boiler is full of water (Fig. 6). Suppose that the water is removed by means of a pump. We wish to find the work done in pumping out the water.

We recall that the work done in moving a particle of matter is the product of the force acting on the particle and the distance through which the particle travels. Returning to our problem, we note that in order to pump a particle of liquid from the boiler it is necessary only to lift it to the brim of the boiler, since from that point it flows out of the boiler under the force of its own weight. Hence the problem reduces to calculating the work needed to lift successively all the particles of the liquid to the rim of the boiler.

Now it is clear that in being lifted to the rim, each particle travels a distance equal to its depth in the boiler. Since the force necessary to lift a particle is the *weight* of the particle, the work done in lifting one particle is equal to the product of its weight and the depth of its submersion in the boiler. Since we are dealing with water, the specific gravity of which is 1, the weight of a given particle is numerically equal to its volume; therefore, *the work done in lifting a particle of water is numerically equal to the product of its volume and the depth of its submersion.*

Since different particles of liquid are to be found at different depths in the boiler, we cannot apply directly the rule just established for calculating the work done. In order to use this rule, we must employ a method of computation similar to that which we adopted for the solution of the problems given in the previous chapter. In other words, we divide the altitude $H$ of the cylinder (Fig. 6) into $n$ parts, each of length $\frac{1}{n} H$, and we draw through the points of division planes parallel to the base of the cylinder. These planes divide the total volume of the liquid into $n$ layers. We can say that within the boundaries of any one layer all the particles of liquid are at approximately the same depth. Therefore, by making use of the rule given above, we can calculate the work done in lifting one layer.

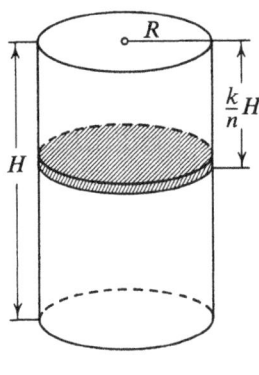

Fig. 6

The volume of a layer is the volume of a cylinder of radius $R$ (where $R$ is the radius of the boiler) and of altitude $\frac{1}{n} H$, which is equal to

$$\pi R^2 \frac{H}{n}.$$

If we consider the $k$th layer from the top, then, clearly, the depth of immersion of this layer is $\frac{k}{n} H$, and thus the work needed to lift the $k$th layer is given by

$$T_k \approx \pi R^2 H^2 \frac{k}{n^2}.$$

The total work $T$ is found by summing the expressions $T_k$, that is,

$$T \approx \sum_{k=1}^{n} \pi R^2 H^2 \frac{k}{n^2} \quad \text{or} \quad T \approx \pi R^2 H^2 \cdot \frac{1}{n^2} \sum_{k=1}^{n} k.$$

Using formula (8) we have

$$T \approx \pi R^2 H^2 \cdot \frac{1}{n^2} \cdot \frac{n(n+1)}{2} \quad \text{or} \quad T \approx \frac{\pi R^2 H^2}{2} \left(1 + \frac{1}{n}\right). \quad (17)$$

This, of course, is only approximate, since, even within a single layer, the depth of submersion of individual particles is not the same. It is clear that, by increasing the number $n$, we can make this approximation increasingly accurate. Hence we get an exact expression for the work done by finding the *limit* of the right-hand side of equality (17) for an infinitely increasing $n$. This limit is found by simply discarding the fraction $\frac{1}{n}$, which gives, finally,

$$T = \frac{\pi R^2 H^2}{2}.$$

Using the expression $V = \pi R^2 H$ for the volume of the cylinder, we can express the total work by

$$T = V\frac{H}{2}.$$

In other words, the work done in which we are interested is equal to the work necessary to lift the contents of the entire boiler a distance of half its altitude.

*Remark.* This last statement can also be established using our method of approximations, but without the calculations. If we have an odd number of layers, there is a middle layer, and the work done in pumping it out is its volume times $\frac{1}{2}H$ (its depth). To each other layer (different from the middle one) there corresponds another layer with the same volume which is at the same distance from the middle layer. If this distance is $d$, one layer must be lifted $\frac{1}{2}H + d$ and the one on the other side $\frac{1}{2}H - d$. The work done in lifting both layers is approximately

$$V'\left(\frac{1}{2}H + d\right) + V'\left(\frac{1}{2}H - d\right) = V'H,$$

where $V'$ is the volume of either layer. Thus, we could assume that both layers were at a depth of $\frac{1}{2}H$, or that all the water was at a depth of $\frac{1}{2}H$. Assuming this, the work done is clearly $\frac{1}{2}VH$. If there were an even number of layers, this result could have been arrived at in a similar manner.

## 15. PUMPING WATER OUT OF A CONICAL TANK

As a second example, let us consider the problem of finding the work done in pumping water out of a conical tank (Fig. 7).

As before, let us divide the whole mass of water into $n$ layers, each $\dfrac{1}{n} H$ thick. The work done in pumping out any layer is equal numerically to the depth of immersion of the layer multiplied by its volume. This volume is the volume of a truncated cone. However, the volume is more readily calculated if the layer is taken to be a cylinder. As we know, this is only an approximation, but just as in section 11, we see that as $n$ increases, the resulting error in this approximation vanishes.

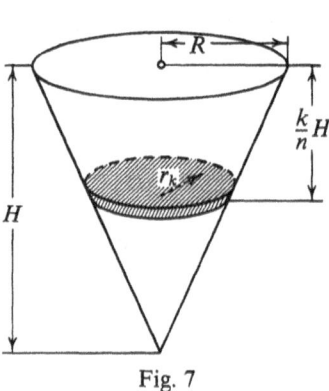

Fig. 7

Denoting the radius of the $k$th layer from the top by $r_k$, we find the volume of this layer to be

$$\pi r_k{}^2 \frac{1}{n} H.$$

Since the depth of submersion of the circular layer is $\dfrac{k}{n} H$, the work required to lift it is equal to

$$T_k \approx \pi r_k{}^2 H^2 \frac{k}{n^2}.$$

The quantity $r_k$ appears in this expression; let us express this quantity in terms of the elements of the cone. From similar triangles we have

$$r_k : R = \left(H - \frac{k}{n} H\right) : H,$$

from which we get

$$r_k = \left(1 - \frac{k}{n}\right) R.$$

Substituting this into the expression for the work required, we find

$$T_k \approx \pi R^2 H^2 \left(1 - \frac{k}{n}\right)^2 \frac{k}{n^2}.$$

The total work necessary, which we are trying to find, is equal to the sum of these terms, that is,

$$T \approx \sum_{k=1}^{n} \pi R^2 H^2 \left(1 - \frac{k}{n}\right)^2 \frac{k}{n^2}$$

or

$$T \approx \pi R^2 H^2 \left[\frac{1}{n^2} \sum_{k=1}^{n} k - \frac{2}{n^3} \sum_{k=1}^{n} k^2 + \frac{1}{n^4} \sum_{k=1}^{n} k^3\right].$$

Using formulas (8), (9), and (10), the latter expression becomes

$$T \approx \pi R^2 H^2 \left[\frac{1}{2}\left(1 + \frac{1}{n}\right) - \frac{1}{3}\left(1 + \frac{1}{n}\right)\left(2 + \frac{1}{n}\right) + \frac{1}{4}\left(1 + \frac{1}{n}\right)^2\right].$$

This expression is only approximate since the layers are not exactly cylindrical and the depths of submersion of the various points of each layer are different. However, if we increase $n$ indefinitely and take the *limit* of the right-hand side, we get the exact value for the work done,

$$T = \pi R^2 H^2 \left(\frac{1}{2} - \frac{2}{3} + \frac{1}{4}\right), \text{ and, finally, } T = \frac{1}{12} \pi R^2 H^2.$$

If we remember[1] that the volume of the cone is equal to

$$V = \frac{1}{3} \pi R^2 H,$$

then the expression we have found for the total work required can be put in the form

$$T = V \cdot \frac{H}{4},$$

which proves to be equal to the work necessary to lift the contents of the entire cone a distance equal to one quarter of its altitude.

[1] This formula, by the way, is proved in section 18.

## 16. PUMPING WATER OUT OF A HEMISPHERICAL TANK

Another problem of the same kind is the computation of the work necessary to pump the water out of a tank having the shape of a hemisphere (Fig. 8).

Proceeding as above, we divide the entire mass of water into $n$ horizontal layers each of thickness $\frac{1}{n} R$. Taking the $k$th layer to be a *cylinder* of radius $r_k$, we see that its volume is

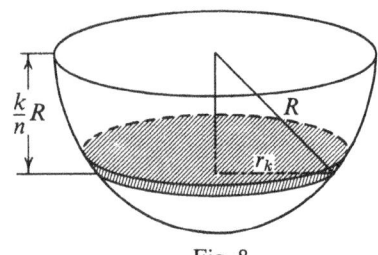

$$V_k \approx \pi r_k^2 \frac{1}{n} R,$$

and, consequently, the work required is

$$T_k \approx \pi r_k^2 \frac{k}{n^2} R^2.$$

Fig. 8

Let us now express the radius $r_k$ of the $k$th layer in terms of the radius $R$ of the sphere. It is not difficult to see from the sketch that we can use the Pythagorean theorem, getting

$$r_k^2 = R^2 - \left(\frac{k}{n} R\right)^2$$

Thus,

$$T_k \approx \pi R^4 \left(1 - \frac{k^2}{n^2}\right) \frac{k}{n^2}.$$

The total work necessary is found by summing all the quantities $T_k$:

$$T \approx \sum_{k=1}^{n} \pi R^4 \left(1 - \frac{k^2}{n^2}\right) \frac{k}{n^2}$$

or

$$T \approx \pi R^4 \left[\frac{1}{n^2} \sum_{k=1}^{n} k - \frac{1}{n^4} \sum_{k=1}^{n} k^3\right].$$

Using the formulas of Chapter 1, we have

$$T \approx \pi R^4 \left[\frac{1}{n^2} \cdot \frac{n(n+1)}{2} - \frac{1}{n^4} \cdot \frac{n^2(n+1)^2}{4}\right],$$

or
$$T \approx \pi R^4 \left[ \frac{1}{2}\left(1 + \frac{1}{n}\right) - \frac{1}{4}\left(1 + \frac{1}{n}\right)^2 \right].$$

This approximate expression for the work required becomes exact if we discard $\frac{1}{n}$, which gives

$$T = \pi R^4 \left(\frac{1}{2} - \frac{1}{4}\right)$$

and, finally,

$$T = \frac{1}{4}\pi R^4.$$

### 17. PUMPING WATER OUT OF A SEMICYLINDRICAL TROUGH

In conclusion, let us consider the problem of calculating the work done in pumping water out of a trough, or a tank, having a semicylindrical shape (Fig. 9).

Applying the same method of division into infinitesimally small quantities, we divide the total mass of water into $n$ narrow

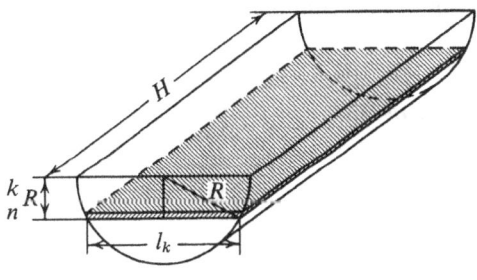

Fig. 9

horizontal layers, each having the shape of a rectangular slab (Fig. 9). The volume of one such slab is equal to

$$V_k \approx H l_k \frac{R}{n},$$

where $l_k$ denotes its width. By the Pythagorean theorem, this width $l_k$ (since it is the chord of a circle at distance $\frac{k}{n}R$ from the center) is

$$l_k = 2 \sqrt{R^2 - \left(\frac{k}{n}R\right)^2};$$

thus the volume of the slab is

$$V_k \approx 2R^2H \frac{1}{n^2} \sqrt{n^2 - k^2},$$

from which it follows that the work of pumping the water out is

$$T_k \approx 2R^3H \frac{k}{n^3} \sqrt{n^2 - k^2}.$$

The total work required is

$$T \approx \sum_{k=1}^{n} 2R^3H \frac{k}{n^3} \sqrt{n^2 - k^2},$$

or

$$T \approx 2R^3H \frac{1}{n^3} \sum_{k=1}^{n} k \sqrt{n^2 - k^2}. \tag{18}$$

However, the expression we have found is only approximate. In order to find the exact value of the work required it is necessary to increase $n$ infinitely and find the *limit* of the right-hand expression in (18).

$$T = 2R^3H \cdot \lim \left[ \frac{1}{n^3} \sum_{k=1}^{n} k \sqrt{n^2 - k^2} \right]. \tag{19}$$

In this way our problem is reduced to finding

$$\lim \left[ \frac{1}{n^3} \sum_{k=1}^{n} k \sqrt{n^2 - k^2} \right] \tag{20}$$

for an infinitely increasing $n$. If we turn to section 13, we see that this limit is the same as the limit (16). We do not know how to find this limit as yet, and thus, we have already found two different physical problems which we are unable to solve completely.

As has already been pointed out in section 13, later on in section 23 we shall find the limit (20), thereby solving both problems.

# 4. Finding Volumes

## 18. THE VOLUME OF A CONE

The methods developed above can be applied to solving a wide range of geometrical problems. In this chapter we shall show how these methods can be employed for finding the volumes of various solids.[1]

Let us consider first of all the problem of finding the *volume of a cone*. To solve this problem let us divide the altitude of the cone into $n$ parts each of height $\frac{1}{n}H$ and through the points of division draw planes parallel to the base of the cone (Fig. 10). These planes divide the entire cone into $n$ layers. Let us, by way of approximation, take each of these layers (which, in fact, are truncated cones) to be a cylinder. This, of course, is not exact, but for large values of $n$ the error is negligible.

Denoting the radius of the $k$th cylinder from the top by $r_k$, we find that the volume of this cylinder is

$$V_k \approx \pi r_k{}^2 \frac{H}{n}.$$

Using the similarity of the triangles, we get

$$r_k : R = \frac{k}{n}H : H,$$

from which we get

$$r_k \approx \frac{k}{n}R,$$

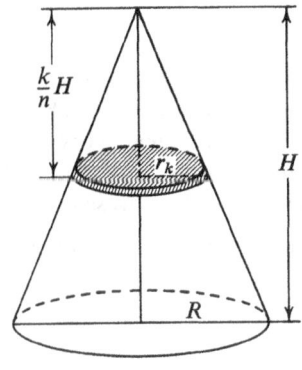

Fig. 10

---

[1] Since we are mainly interested in the calculation of volumes, we do not consider here the question of a precise *definition* of the concept "volume." However, for such a definition we still need the concept of limit.

and the expression for the volume takes the form

$$V_k \approx \pi R^2 H \frac{k^2}{n^3}.$$

The total volume is

$$V \approx \sum_{k=1}^{n} \pi R^2 H \frac{k^2}{n^3},$$

or

$$V \approx \pi R^2 H \cdot \frac{1}{n^3} \sum_{k=1}^{n} k^2,$$

which, on the basis of formula (9), is equal to

$$V \approx \pi R^2 H \frac{n(n+1)(2n+1)}{6n^3},$$

or

$$V \approx \pi R^2 H \frac{\left(1 + \dfrac{1}{n}\right)\left(2 + \dfrac{1}{n}\right)}{6}. \tag{21}$$

This expression for the volume is not exact but only approximate because, as has been pointed out, the individual layers are not really cylinders. However, as $n$ increases, this expression becomes increasingly accurate, so that the true value of $V$ is the *limit* of the right-hand expression of (21) for an infinitely increasing $n$. This limit can be found from (21) by discarding the fractions $\frac{1}{n}$. Thus,

$$V = \pi R^2 H \cdot \frac{1 \cdot 2}{6}$$

and, finally,

$$V = \frac{1}{3} \pi R^2 H.$$

Hence, *the volume of a cone is equal to one third of the product of the area of its base and its altitude.*[1]

[1] The reader should note that, although the cone pictured in Fig. 10 is a *right* circular cone, the derivation of the formula for the volume is also valid for *oblique* circular cones (that is, cones whose altitude and axis fail to coincide).

## 19. THE VOLUME OF A PYRAMID

Similar reasoning enables us to find the volume of a pyramid. Let us consider a pyramid of altitude $H$, the area of whose base is $F$ (Fig. 11). Dividing the altitude into $n$ equal parts and drawing through the points of division planes parallel to the base, we cut the pyramid into $n$ prismatic slabs each of height $\frac{1}{n}H$. (Strictly

speaking, these slabs are not prisms but are truncated pyramids; but, as above, we can, by way of approximation, take them to be prismatic.)

If the area of the $k$th slab from the top is $F_k$, then it is not difficult to see that the proportion[1]

$$F_k : F = k^2 : n^2$$

holds true, so that

$$F_k = \frac{k^2}{n^2} F,$$

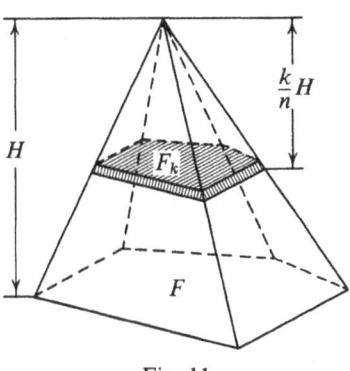

Fig. 11

and, hence, the volume of the $k$th slab is

$$V_k \approx F_k \cdot \frac{H}{n} = \frac{k^2}{n^3} FH.$$

The volume of the entire pyramid is equal to the sum of these volumes.

$$V \approx \frac{FH}{n^3} \sum_{k=1}^{n} k^2, \text{ or, from formula (9)}, V \approx \frac{FH}{6}\left(1 + \frac{1}{n}\right)\left(2 + \frac{1}{n}\right).$$

Making $n$ indefinitely large and taking the limit of the right-hand side, we find the exact expression

$$V = \frac{1}{3} FH;$$

*the volume of a pyramid is equal to one third the product of the area of its base and its altitude.* This is analogous to the result for a cone.

[1] To show that this proportion holds, first show that it holds if the base is a triangle or a rectangle. Once the ratio is established in this special case, it can easily be shown to hold if the base can be decomposed into a finite number of triangles.

## 20. THE VOLUME OF A SPHERE

Now let us find the volume of a sphere. Evidently, the problem will be solved if we restrict ourselves to a *hemisphere $V^*$* and then double the result. Dividing the hemisphere (Fig. 12) by means of planes into $n$ layers, each of thickness $\frac{1}{n} R$, we begin by assuming

these layers to be cylinders. If the radius of the $k$th layer is $r_k$, then its volume, being approximately the volume of a cylinder, is

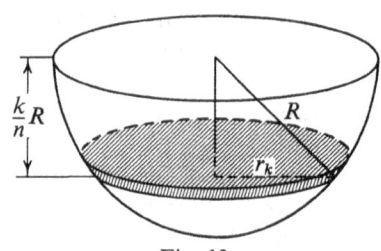

$$V_k \approx \pi r_k^2 \frac{R}{n}.$$

The Pythagorean theorem gives

$$r_k^2 = R^2 - \frac{k^2}{n^2} R^2,$$

Fig. 12

so that the expression for the volume takes the form

$$V_k \approx \pi R^3 \left( 1 - \frac{k^2}{n^2} \right) \frac{1}{n}.$$

But the volume $V^*$ of the entire hemisphere is the sum of all the volumes $V_k$:

$$V^* \approx \pi R^3 \left[ \sum_{k=1}^{n} \frac{1}{n} - \frac{1}{n^3} \sum_{k=1}^{n} k^2 \right],$$

which, on the basis of the properties shown in Chapter 1, is

$$V^* \approx \pi R^3 \cdot \frac{6 - \left( 1 + \frac{1}{n} \right)\left( 2 + \frac{1}{n} \right)}{6}.$$

The limit of this expression, for an infinitely increasing $n$, gives as the exact value of the volume of the hemisphere

$$V^* = \frac{2}{3} \pi R^3;$$

thus, the volume of the entire sphere is

$$V = \frac{4}{3} \pi R^3.$$

## 21. THE VOLUME OF THE COMMON PORTION OF TWO CYLINDERS

Now let us solve a more difficult problem. Let us consider two cylinders having equal radii, the axes of which intersect at right angles (Fig. 13). We pose the problem of finding the volume of the solid *which is the portion common to both cylinders*. The complicated nature of this solid and the consequent difficulty of finding an exact expression for it make the problem difficult.

For this purpose let us imagine a plane passing through the axes of both cylinders; let us call it the "axial" plane. This plane (if we think of it as coinciding with the plane of the drawing) divides the solid into two equal halves: "front" and "back." Let us restrict ourselves to one of these, the front one for example, since the two halves are clearly congruent.

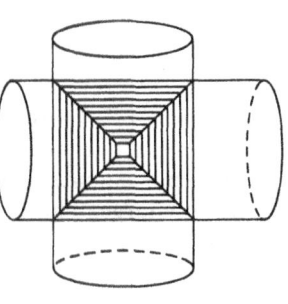

Now let us consider any plane parallel to the axial plane. It cuts each of the cylinders in a strip; it is clear, moreover, that both strips have the same width. Hence, the solid we are considering intersects the plane in a *square*. Having established this, it is not difficult to solve our problem. First we draw a perpendicular to the axial plane from the point of intersection of the axes of the cylinders. The length of the segment it cuts off in the front half of the solid in which we are interested is equal to

Fig. 13

$R$. Divide this segment into $n$ parts and draw through the points of division planes parallel to the axial plane. These planes divide the front half of the solid we are considering into $n$ square slabs each

of thickness $\frac{1}{n} R$.

It is not difficult to see from Fig. 14, in which the solid we are considering is pictured from above, that the side of the $k$th square is equal to

$$l_k = 2 \sqrt{\left(R^2 - \frac{k}{n} R\right)^2},$$

so that its area is $l_k{}^2 = 4R^2 \left(1 - \frac{k^2}{n^2}\right)$.

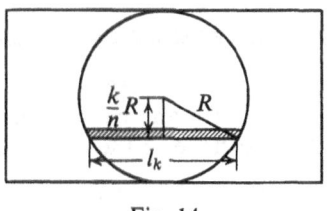

Fig. 14

Consequently, the volume of the $k$th slab is

$$V_k \approx l_k{}^2 \cdot \frac{R}{n} = 4R^3 \left(1 - \frac{k^2}{n^2}\right)\frac{1}{n}.$$

The volume $V^*$ of the entire front half of the solid is the sum of all the $V_k$, that is,

$$V^* \approx \sum_{k=1}^{n} 4R^3 \left(1 - \frac{k^2}{n^2}\right)\frac{1}{n},$$

so that

$$V^* \approx 4R^3 \left[ \sum_{k=1}^{n} \frac{1}{n} - \frac{1}{n^3} \sum_{k=1}^{n} k^2 \right],$$

or

$$V^* \approx 4R^3 \left[ 1 - \frac{1}{6}\left(1 + \frac{1}{n}\right)\left(2 + \frac{1}{n}\right) \right].$$

This approximate expression leads to an exact one by making $n$ indefinitely large. Thus, the volume of the front half of the solid is given by

$$V^* = \frac{8}{3} R^3.$$

The total volume $V$ is obtained by doubling this value, that is,

$$V = \frac{16}{3} R^3;$$

this completes the solution of the problem. It is worth noting that, in spite of the rather complicated nature of the solid, its volume can be expressed without using any irrational numbers.[1]

## 22. THE VOLUME OF A CYLINDRICAL SEGMENT

A *cylindrical segment* is a solid bounded by the base of a given cylinder, a plane passing through a diameter of this base, and the lateral surface of the cylinder (Fig. 15). Using the sketch in Fig. 15,

[1] A rational number is one which can be expressed as a quotient $\frac{p}{q}$ of two whole numbers ($q \neq 0$).

let $|AB| = H$, $|OA| = R$.[1] We seek a formula that will give the volume of the segment in terms of $H$ and $R$.

We begin by deriving a formula which yields a good *approximate* value for the volume of the cylindrical segment. To do this, let us cut the radius $OK$ into $n$ equal line segments and through the points of division let us draw planes parallel to the plane of triangle $OAB$. These planes divide the half of the cylindrical segment (the half determined by $OK$) into $n$ slabs. Consider the slab determined by the triangles $O_1A_1B_1$ and $O_2A_2B_2$ (see Fig. 15). Its thickness is the same as that of all the other slabs—specifically, $\frac{1}{n}R$; its volume

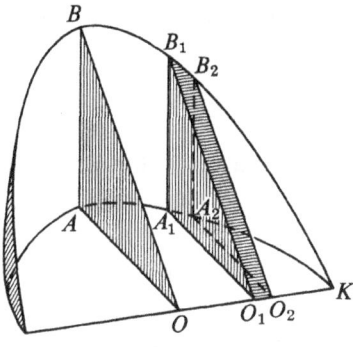

Fig. 15

is approximately equal to the volume of a thin right triangular prism having the triangle $O_1A_1B_1$ for its base and $O_1O_2$ as a lateral edge. Let us find the volume of this prism, assuming for purposes of calculation that the slab being considered is the $k$th slab.

Since $O_1A_1B_1B_2A_2O_2$ is the $k$th slab, we have

$$|OO_2| = \frac{k}{n}R.$$

Applying the Pythagorean theorem to the right triangle $OO_2A_2$, we obtain
$$|O_2A_2| = \sqrt{|OA_2|^2 - |OO_2|^2}.$$

Noting that $|OA_2| = R$, making the substitutions for this and $|OO_2|$ in the last equality, and simplifying the result, we obtain

$$|O_2A_2| = R \sqrt{1 - \frac{k^2}{n^2}}.$$

Furthermore, from the similarity of the triangles $OAB$ and $O_2A_2B_2$ it follows that

$$|A_2B_2|:|AB| = |O_2A_2|:|OA| \quad \text{or} \quad |A_2B_2|:H = R\sqrt{1 - \frac{k^2}{n^2}} : R,$$

[1] Here, as well as in later sections, we use the symbol "$|AB|$" to represent the *length* of the straight line segment $AB$.

so that
$$|A_2B_2| = H\sqrt{1 - \frac{k^2}{n^2}}.$$

It is now easy to calculate the volume of our prism. The base of the prism is triangle $O_2A_2B_2$; the area of triangle $O_2A_2B_2$ is $\frac{1}{2}|O_2A_2| \cdot |A_2B_2|$, or, using the results obtained above,

$$\frac{1}{2}RH\left(1 - \frac{k^2}{n^2}\right).$$

The altitude of the prism is equal to the thickness of the slab, $\frac{1}{n}R$. Thus, the volume of the prism is

$$\frac{1}{2}R^2H\left(\frac{1}{n} - \frac{k^2}{n^3}\right).$$

We have already noted that the volume of the $k$th slab is approximately the volume of the $k$th prism. Thus we may write

$$V_k \approx \frac{1}{2}R^2H\left(\frac{1}{n} - \frac{k^2}{n^3}\right),$$

where $V_k$ is the volume of the $k$th slab.

By adding the volumes of the prisms, we obtain an approximation for the volume, $V^*$, of the entire half of the original cylindrical segment:

$$V^* \approx \frac{1}{2}R^2H\left[\sum_{k=1}^{n}\frac{1}{n} - \sum_{k=1}^{n}\frac{k^2}{n^3}\right] \text{ or } V^* \approx \frac{1}{2}R^2H\left[1 - \frac{\left(1 + \frac{1}{n}\right)\left(2 + \frac{1}{n}\right)}{6}\right].$$

As $n$ grows larger our approximation becomes more and more accurate, and if we pass on to the limit of the expression on the right of the latter result, we easily obtain the desired formula. Taking the limit of this expression, we have

$$V^* = \frac{1}{3}R^2H;$$

hence the volume $V$ of the entire cylindrical segment is $2V^*$:

$$V = \frac{2}{3}R^2H. \tag{22}$$

34

## 23. ANOTHER METHOD OF SOLUTION; SOLUTIONS OF PROBLEMS OF SECTIONS 13 AND 17

Let us try to solve the preceding problem by another method. Let us divide the radius $OA$ into $n$ parts (Fig. 16) and draw through the points of division planes perpendicular to this radius. The entire cylindrical segment is divided into $n$ rectangular slabs.

Let us consider the volume of the $k$th slab. (We shall assume the slabs to be prismatic.) The thickness of each slab is $\frac{1}{n}R$, so that

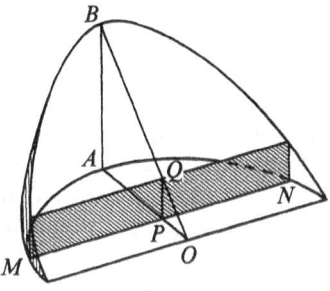

the problem reduces to finding the area of the base of the slab. Assuming that we are talking about the $k$th slab, we have

$$|OP| = \frac{k}{n} R.$$

In that case, by the Pythagorean theorem, the length of the chord $MN$ is

$$|MN| = 2 \sqrt{R^2 - \frac{k^2}{n^2} R^2}.$$

Fig. 16

Using the similarity of the triangles $OPQ$ and $OAB$, we have

$$|PQ| : |OP| = |AB| : |OA|, \text{ or } |PQ| : \frac{k}{n} R = H : R,$$

so that

$$|PQ| = \frac{k}{n} H.$$

The area of the rectangle is $|PQ| \cdot |MN|$; substitution of the expressions found for $|PQ|$ and $|MN|$ yields

$$2RH \frac{k}{n} \sqrt{1 - \frac{k^2}{n^2}}.$$

Thus, multiplying by the thickness of the slab, we get the volume as

$$V_k \approx 2R^2H \cdot \frac{k}{n^3} \sqrt{n^2 - k^2}.$$

Therefore, the total volume is equal to the sum

$$V \approx 2R^2H \cdot \frac{1}{n^3} \sum_{k=1}^{n} k \sqrt{n^2 - k^2}.$$

35

However, this expression is only an approximation; from it we get the *exact* expression for the volume if we replace the right-hand side of the preceding expression by its *limit* for infinitely increasing $n$. This gives

$$V = 2R^2H \lim \left[\frac{1}{n^3} \sum_{k=1}^{n} k \sqrt{n^2 - k^2}\right]. \tag{23}$$

Here, for the third time, we encounter

$$\lim \left[\frac{1}{n^3} \sum_{k=1}^{n} k \sqrt{n^2 - k^2}\right].$$

We still have no *direct* means for calculating such a limit. But in the preceding section we derived a formula for the volume $V$. If we compare expressions (22) and (23), we can determine the value of the limit in which we are interested. Equating the right-hand side of (22) to the right-hand side of (23), we obtain

$$\frac{2}{3} R^2H = 2R^2H \lim \left[\frac{1}{n^3} \sum_{k=1}^{n} k \sqrt{n^2 - k^2}\right]$$

or

$$(2R^2H) \cdot \frac{1}{3} = 2R^2H \lim \left[\frac{1}{n^3} \sum_{k=1}^{n} k \sqrt{n^2 - k^2}\right].$$

Dividing by $2R^2H$, we find at once that

$$\lim \left[\frac{1}{n^3} \sum_{k=1}^{n} k \sqrt{n^2 - k^2}\right] = \frac{1}{3}. \tag{24}$$

Hence, we have finally determined this limit.

Returning to section 13, let us substitute the limit we have found into equality (15) and at once obtain the value of the pressure we were seeking:

$$P = \frac{2}{3} R^3.$$

In the same way if we insert this limit into equality (19) in section 17, we find the work required, which we were seeking, to be

$$T = \frac{2}{3} R^3H.$$

## 24. GENERAL REMARKS

All the problems posed above have been solved by essentially the same method. This method consists of the following: The quantity to be determined is expressed in the form of a sum of a large number of very small terms of the same kind. These small terms are calculated approximately, but with such care that, by increasing the number of terms, this sum approximates the precise value more and more closely. Nevertheless, the sum which is obtained for the quantity sought turns out to be inexact. In order to find the exact value, we are obliged to consider the *limit* of the sum as the number of terms increases indefinitely.

In short, the method outlined consists of representing the quantity sought in the form of *the limit of the sum of an infinitely increasing number of infinitely decreasing terms,* or, as it is often expressed, in the form of *the sum of an infinitely large number of infinitely small quantities.*[1]

This method is one of the most important methods of higher mathematics; it is studied in that branch known as the *integral calculus.* There a study is made precisely of the limits of sums of an infinitely increasing number of infinitely decreasing terms. These limits are called *integrals.* Thus, regarding our solutions of the problems of the preceding paragraphs, we can say that in each of them we have been computing certain integrals.

The sums which we examined had very simple forms. Actually, many of our sums had one of the following three forms:

$$\sum_{k=1}^{n} k, \qquad \sum_{k=1}^{n} k^2, \qquad \sum_{k=1}^{n} k^3.$$

In calculating each sum we referred to the appropriate formula of Chapter 1. When we encountered a sum of more complicated form, namely, the sum

$$\frac{1}{n^3} \sum_{k=1}^{n} k \sqrt{n^2 - k^2},$$

then only by resorting to somewhat artificial reasoning were we able to calculate the desired limit. Indeed, had we not hit on the fortunate idea of solving the problem of section 22 by two methods, we

[1] This expression is not to be taken literally.

would not have found this limit and would not have solved the problems of sections 13 and 17. *General methods* for finding the limits of sums, often even of very complicated forms, are explained in integral calculus, so that the solution of a large class of similar problems is greatly simplified and becomes "mechanical."

Mathematicians did not discover these general methods all at once. On the contrary, their discovery was the result of the collective work of many generations. These methods assumed their contemporary form in the works of Leibniz (1646–1716) and Newton (1642–1727); however, the idea of dividing into an infinitely large number of infinitely small terms was known much earlier. Strictly speaking, this idea was familiar to the mathematicians of ancient Greece (in a general form to Archimedes, 287–212 B.C.). In particular, Archimedes knew how to compute the volume of a sphere, a cone, portions of them, and even of a cylindrical segment.

During the period of the Middle Ages, scientific thought was in a condition of deep decline, and it was only with the beginning of the sixteenth century that the sciences (and, in particular, mathematics) began to be developed once more. At first scientists merely rediscovered the results of ancient times; but subsequently, they gradually began to progress beyond the Greeks. This was true in the case of the method of summation of infinitely small quantities, with which we are concerned. This method received a large stimulus in the works of Kepler's *Solid Geometry of Wine Casks* (1615) and Cavalieri's *Geometry of Indivisibles* (1635).

However, these last two authors did not, as yet, have a *general* method for finding the limits of sums. Thus, the explanation which is given in our booklet approximates, in its scope, that of the works of Kepler and Cavalieri (differing from them mainly in exposition).

Gradually, in later research, more and more general methods were found for calculating integrals. As has been pointed out already, this problem was finally solved in a completely general way by Leibniz and Newton. (The very term "integral" belongs to the school of Leibniz and was introduced in the year 1690.)

## 25. THE PRINCIPLE OF CAVALIERI

Cavalieri was unable to find the limits of sums of complicated form; however, he discovered a very useful principle which helped him to avoid calculating these sums in many cases. This principle is formulated in the following way.

*If two solids are contained between two parallel planes P and Q (Fig. 17) and possess the property that the sections of them, cut by any plane R parallel to P and Q, always are of equal area, then the volumes of these solids are equal.*

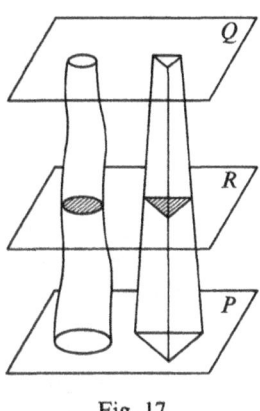

In order to establish this principle, let us draw $n - 1$ equally spaced planes between and parallel to $P$ and $Q$. These planes divide the solids into $n$ layers. If we assume that these layers are approximately cylindrical or prismatic in shape, then we find that their volumes are equal. Therefore, the volumes of the original solids are equal also, since they are both found by summing the volumes of the layers. This equality of volumes seems, at first, to be merely approximate, but since it may be found to any degree of accuracy, we can be sure that it is absolutely accurate.

Fig. 17

It is not difficult to generalize this principle showing that if the areas of the two sections described above bear a constant ratio to one another, then the ratio of their volumes is the same constant. It is not difficult to establish a similar principle for areas as well.

The formulation of the principle in this case is as follows:

*If two plane figures I and II, contained between parallel lines p and q (Fig. 18) have the property that sections of them formed by any line r, parallel to p and q, have equal lengths, then the figures have the same area.*

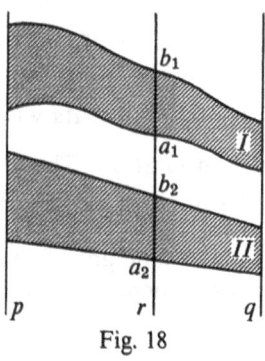

If the ratio of the lengths of the segments $a_1b_1$ and $a_2b_2$ is equal to $k$, regardless of the position of $r$, then the ratio of the area of *I* to the area of *II* is equal to $k$.

The proofs of these statements are analogous to those given in the case of volumes and are left to the reader.

Fig. 18

# 5. The Parabola and the Ellipse

## 26. THE AREA UNDER A PARABOLA

Let us consider a curve whose equation in a rectangular coordinate system is

$$y = ax^2. \qquad (25)$$

Such a curve is called a *parabola;* it has the shape shown in Fig. 19 (we assume that $a > 0$).

Taking any arbitrary point $M$ on the parabola, let us drop a perpendicular $MP$ from $M$ to the $x$-axis. Let us pose the problem of finding the area $F$ of the curvilinear triangle $OMP$.

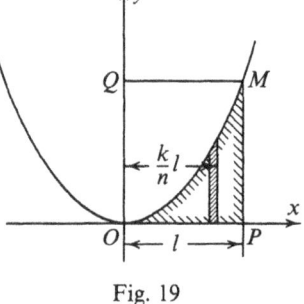

In order to solve this problem, we divide the segment $OP$ into $n$ equal parts, and we draw from the points of division perpendiculars intersecting the parabola. These perpendiculars divide the region whose area we are seeking into $n$ narrow vertical strips. These

Fig. 19

strips may be considered as approximately rectangular. Let us compute their areas on this supposition.

Let us denote the length of the segment $OP$ by $l$ and consider the $k$th strip. Its width is $\frac{1}{n} l$. Its height is found by the following computation: The distance of the strip from the $y$-axis is equal to $\frac{k}{n} l$ and, since one of the upper vertices of the strip lies on the parabola, the height of the strip is equal to the $y$-coordinate of this vertex, which, according to equation (25), is equal to

$$a \left( \frac{k}{n} l \right)^2.$$

Hence, the area of the strip is

$$al^3 \frac{k^2}{n^3},$$

so that the area of the whole region $OMP$ is approximately the sum

$$F \approx \sum_{k=1}^{n} al^3 \frac{k^2}{n^3} \text{ or } F \approx al^3 \cdot \frac{1}{n^3} \sum_{k=1}^{n} k^2$$

or, finally,

$$F \approx \frac{al^3}{6}\left(1 + \frac{1}{n}\right)\left(2 + \frac{1}{n}\right).$$

In order to obtain from this the exact equality, it is necessary to increase the number $n$ infinitely. For the limit we get

$$F = \frac{al^3}{3}.$$

This result can be given a simple geometrical formulation. In fact, let us consider the *rectangle OQMP*. Its area is, evidently, equal to $|OP| \cdot |PM|$. But $|OP| = l$; the length of $PM$ is the ordinate of the point $M$, the abscissa (first coordinate) of which is $l$, so that it follows from the equation of the parabola that $|PM| = al^2$.

Thus, the area of $OQMP$ is $al^3$ and, consequently, *the area of OMP is equal to one third the area of the rectangle OQMP.* Hence, the area of $OQM$ is equal to two thirds the area of the same rectangle. These elegant results were first discovered by Archimedes.

In general, finding the area of a region is called the *quadrature* of this region (since it consists in comparing the area of the region to the area of a *square*). Thus, we have effected the quadrature of the parabola.

### 27. THE VOLUME OF A PARABOLOID OF REVOLUTION

Let us suppose that the parabola considered in the previous section is revolved about the $y$-axis (Fig. 20). The solid obtained in this way is called a *paraboloid of revolution*. Let us consider the plane $A$, perpendicular to the $y$-axis; we shall find the volume of the solid bounded by the paraboloid and this plane.

To do this, we divide the segment $OQ$ (see Fig. 20) into $n$ equal parts and draw through the points of division planes parallel to the plane $A$. These planes divide the solid which we are considering into $n$ layers, each of which we take to be approximately a cylinder. As before, we denote the distance $|OP|$ by $l$; then, as above, we find that $|OQ| = al^2$. Thus the height of each cylinder is $\frac{1}{n} al^2$.

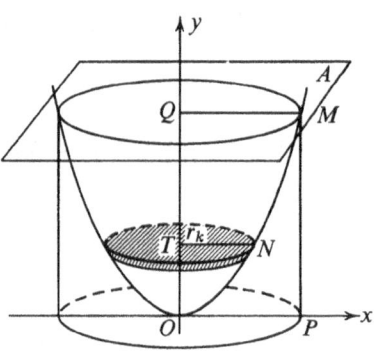

Fig. 20

In order to find the radius of the $k$th cylinder, we proceed in the following manner: the radius $r_k = |NT|$ evidently represents the abscissa of the point $N$ of the parabola. Since the ordinate of this point is

$$\frac{k}{n}|OQ| = \frac{k}{n}al^2,$$

we find, from the equation of the parabola, that

$$\frac{k}{n}al^2 = ar_k^2, \text{ whence } r_k^2 = \frac{k}{n}l^2,$$

so that the area of the base of the $k$th cylinder is

$$\pi r_k^2 = \pi l^2 \frac{k}{n}.$$

Thus, the volume of the $k$th cylinder is

$$V_k \approx a\pi l^4 \frac{k}{n^2}.$$

This means that the total volume $V$ which we are seeking is

$$V \approx a\pi l^4 \cdot \frac{1}{n^2} \sum_{k=1}^{n} k;$$

this yields, after a simple calculation,

$$V \approx a\pi l^4 \frac{1}{2} \left(1 + \frac{1}{n}\right).$$

If we increase $n$ infinitely, we find the exact expression for the volume of the paraboloid of revolution:

$$V = \frac{1}{2}\pi al^4.$$

Let us compare this volume with the volume of a cylinder of radius $R = |OP|$ and altitude $H = |OQ|$. Its volume is

$$\pi R^2 H = \pi |OP|^2 \cdot |OQ| = \pi l^2 \cdot al^2 = \pi al^4.$$

Thus we have the THEOREM OF ARCHIMEDES:

*The volume of a paraboloid of revolution is equal to half the volume of a cylinder of the same base and altitude.*

## 28. THE ELLIPSE AND ITS AREA

We now consider a very important curve called an *ellipse*. The definition of this curve is: An ellipse is a contracted circle. Let us elucidate this definition.

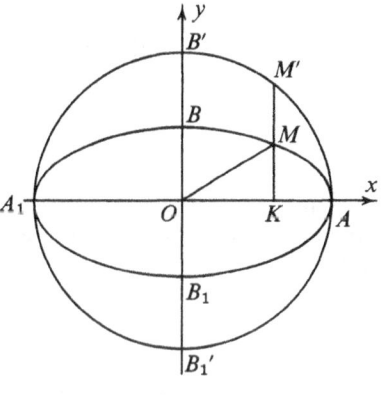

Let us consider a circle of radius $a$. Suppose that it lies in a plane with a rectangular coordinate system, and that its center coincides with the origin of the coordinate system (Fig. 21). Suppose, further, that the ordinates $KM'$ of all points $M'$ of the circle are shortened by a coefficient of contraction $q < 1$:

$$KM : KM' = q.$$

This operation of shortening transforms the circle $A_1 B' A B_1'$ into an ellipse. Let us derive the equation of the ellipse.

Fig. 21

If we denote the coordinates of the point $M$ of the ellipse by $x$ and $y$, then we have from the definition of the ellipse that

$$y = q \cdot KM'.$$

By the Pythagorean theorem,

$$|KM'| = \sqrt{(OM')^2 - (OK)^2} = \sqrt{a^2 - x^2}.$$

Depending on whether $M$ lies on the upper or lower half of the ellipse we have

$$KM' = \sqrt{a^2 - x^2} \quad \text{or} \quad KM' = -\sqrt{a^2 - x^2},$$

so that

$$y = q\sqrt{a^2 - x^2} \quad \text{or} \quad y = -q\sqrt{a^2 - x^2}.$$

If we denote the length $|OB|$ by $b$, then from the definition of the ellipse we have

$$b : a = |OB| : |OB'| = q,$$

so that

$$q = \frac{b}{a}.$$

Thus, if $M$ lies on the upper half of the ellipse, its coordinates satisfy the equation

$$y = \frac{b}{a}\sqrt{a^2 - x^2},$$

while if $M$ lies on the lower half of the ellipse, they satisfy the equation

$$y = -\frac{b}{a}\sqrt{a^2 - x^2}.$$

Upon squaring these two equations, we see that regardless of where the point $M$ lies on the ellipse, its coordinates satisfy the equation

$$y^2 = \frac{b^2}{a^2}(a^2 - x^2).$$

Dividing both sides of the equation by $b^2$, we obtain

$$\frac{y^2}{b^2} = \frac{1}{a^2}(a^2 - x^2) = 1 - \frac{x^2}{a^2},$$

or, finally,

$$\frac{x^2}{a^2} + \frac{y^2}{b^2} = 1.$$

This last equation is called the "canonical" or "simplest" equation of the ellipse.

Let us determine the area of an ellipse. If we use the remarks concerning the principle of Cavalieri made at the end of section 25, we can say at once that *the ratio of the area of the ellipse to the area of the circle is equal to the coefficient of contraction q*; thus, denoting the area of the ellipse by $F$, we have

$$F : \pi a^2 = q,$$

or

$$F = q\pi a^2.$$

On substituting $\dfrac{b}{a}$ for $q$ in the latter equality, we finally obtain

$$F = \pi ab.$$

Moreover, if we use the principle of Cavalieri, we can easily find also the volume of an *ellipsoid of revolution,* that is, of the solid formed by rotating an ellipse about the $x$-axis (Fig. 22). To be explicit, the ratio of the radii of the circles found by cutting the ellipsoid sectionally by planes perpendicular to the $x$-axis, to the radii of the sections of the sphere cut by the same planes, is equal to $q$. Thus, the ratio of the areas of these sections is equal to $q^2$. By the principle of Cavalieri, we have

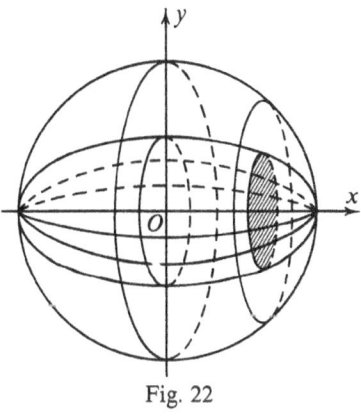

Fig. 22

the ratio of the volumes of the ellipsoid and the sphere as well. Thus,

$$V : \frac{4}{3}\pi a^3 = q^2 = \frac{b^2}{a^2},$$

so that

$$V = \frac{4}{3}\pi ab^2.$$

# 6. The Sinusoid

## 29. A TRIGONOMETRIC SUM

In the presentation which follows, we shall require an expression for the following sum:

$$S = \sum_{k=1}^{n} \sin k\alpha = \sin \alpha + \sin 2\alpha + \cdots + \sin n\alpha, \qquad (26)$$

where $\alpha$ is some definite angle.

To find this sum, let us multiply both sides of equality (26) by $2 \sin \frac{\alpha}{2}$:

$$2S \sin \frac{\alpha}{2} = 2 \sin \alpha \sin \frac{\alpha}{2} + 2 \sin 2\alpha \sin \frac{\alpha}{2} + \cdots$$
$$+ 2 \sin n\alpha \sin \frac{\alpha}{2},$$

and apply to each term of the right-hand side the well-known formula

$$2 \sin A \sin B = \cos (A - B) - \cos (A + B).$$

This gives

$$2S \sin \frac{\alpha}{2} = \left[ \cos \frac{\alpha}{2} - \cos \frac{3}{2} \alpha \right]$$
$$+ \left[ \cos \frac{3}{2} \alpha - \cos \frac{5}{2} \alpha \right]$$
$$+ \cdots$$
$$+ \left[ \cos \frac{2n - 1}{2} \alpha - \cos \frac{2n + 1}{2} \alpha \right].$$

It is easy to see that in each set of brackets (except the first), the first term cancels with the second term of the preceding set of brackets. Thus,

$$2S \sin \frac{\alpha}{2} = \cos \frac{\alpha}{2} - \cos \frac{2n + 1}{2}\alpha. \tag{27}$$

Using the formula

$$\cos A - \cos B = 2 \sin \frac{B - A}{2} \sin \frac{A + B}{2},$$

we can put (27) in the form

$$2S \sin \frac{\alpha}{2} = 2 \sin \frac{n\alpha}{2} \sin \frac{(n + 1)\alpha}{2},$$

from which it follows that

$$S = \frac{\sin \dfrac{n\alpha}{2} \sin \dfrac{(n + 1)\alpha}{2}}{\sin \dfrac{\alpha}{2}}.$$

Therefore,

$$\sum_{k=1}^{n} \sin k\alpha = \frac{\sin \dfrac{n\alpha}{2} \sin \dfrac{(n + 1)\alpha}{2}}{\sin \dfrac{\alpha}{2}}. \tag{28}$$

This is the desired formula.

## 30. A SUBSIDIARY INEQUALITY

Let $\alpha$ denote an arbitrary angle[1] satisfying the condition $0 < \alpha < \frac{\pi}{2}$.[2] In this case we have the following double inequality:

$$\tan \alpha > \alpha > \sin \alpha. \tag{29}$$

In order to prove this statement let us consider Fig. 23. From this figure we immediately see that the triangle $OCA$ is completely

---

[1] More precisely, $\alpha$ is the measure of the angle in radians.

[2] If $A$ and $B$ are two numbers, then $A < B$ and $B > A$ both mean $A$ is less than $B$ or, in other words, $B$ is greater than $A$. The following facts about inequalities are used in the discussion which follows: If $A$, $B$, and $C$ are positive numbers and $A < B$, then $1/A > 1/B$ and $A/C < B/C$.

contained within the sector $OCA$, which, in turn, is completely contained within the triangle $OAB$. Consequently, the following inequalities hold for the areas of these figures:

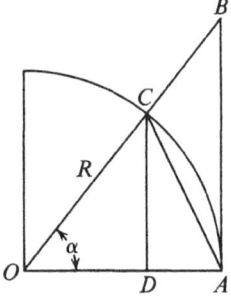

$$\text{area of } \triangle OAB > \text{area of sector } OCA$$
$$> \text{area of } \triangle OCA.$$

In other words,

$$\frac{1}{2}|OA|\cdot|AB| > \frac{1}{2}R\cdot\widehat{CA} > \frac{1}{2}|OA|\cdot|CD|.$$

But

$$|OA| = R, \quad |AB| = R\tan\alpha,$$
$$\widehat{CA} = R\alpha, \quad |CD| = R\sin\alpha,$$

Fig. 23

so that

$$\frac{1}{2}R^2\tan\alpha > \frac{1}{2}R^2\alpha > \frac{1}{2}R^2\sin\alpha.$$

If we divide all members of this double inequality by the positive number $\frac{1}{2}R^2$, we obtain inequality (29).

### 31. THE SINE OF AN INFINITELY SMALL ANGLE

Let us suppose that the angle $\alpha$ tends to zero, successively assuming values $\alpha_1, \alpha_2, \alpha_3, \ldots$. In this case we have the formula

$$\lim \frac{\sin \alpha_n}{\alpha_n} = 1, \tag{30}$$

which is one of the most important formulas of mathematics.

It is helpful to learn formula (30) in words: *The limit of the ratio of the sine of an infinitely small angle to the value of this angle measured in radians is equal to one.*

In order to prove this formula, we can assume that all the values of $\alpha_n$ are positive, since the value of the ratio $\dfrac{\sin \alpha_n}{\alpha_n}$ does not change if $\alpha_n$ is changed to $-\alpha_n$. Moreover, we can assume that $\alpha_n < \dfrac{\pi}{2}$, for, in any event, this will be the case for sufficiently large values of $n$. Hence,

$$0 < \alpha_n < \frac{\pi}{2},$$

and then, by virtue of formula (29),

$$\tan \alpha_n > \alpha_n > \sin \alpha_n;$$

therefore, dividing all terms of the inequality by the positive quantity $\sin \alpha_n$, we obtain

$$\frac{1}{\cos \alpha_n} > \frac{\alpha_n}{\sin \alpha_n} > 1.$$

By inverting the terms of this inequality, we get the inequality

$$\cos \alpha_n < \frac{\sin \alpha_n}{\alpha_n} < 1. \tag{31}$$

It is stipulated that $\alpha_n$ tends to zero. Then (as one may easily see from the sketch) the cosine of this angle tends to 1:

$$\lim (\cos \alpha_n) = 1;$$

but since, in accordance with (31), the fraction $\dfrac{\sin \alpha_n}{\alpha_n}$ lies between 1 and $\cos \alpha_n$, it must tend to 1 also. This proves formula (30).

## 32. THE QUADRATURE OF THE SINUSOID

Let us consider the curve for which the equation is

$$y = \sin x. \tag{32}$$

The curve is sketched in Fig. 24 and is called a *sinusoid*.

Let us find the area of the region bounded by the section of the sinusoid from $x = 0$ to $x = \pi$ and the axis of the abscissa. (This region is shaded in Fig. 24.)

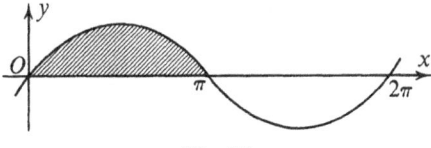

Fig. 24

As usual, we begin by dividing the axis of the abscissa from $x = 0$ to $x = \pi$ into $n$ parts by the points

$$x_1 = \frac{\pi}{n}, \quad x_2 = \frac{2\pi}{n}, \quad \ldots, \quad x_n = \frac{n\pi}{n};$$

from these points we then draw perpendiculars which intersect the sinusoid. From equation (32) we find the lengths of these perpendiculars to be

$$\sin \frac{\pi}{n}, \quad \sin \frac{2\pi}{n}, \quad \sin \frac{3\pi}{n}, \quad \ldots, \quad \sin \frac{n\pi}{n}$$

(so that the last of them is equal to zero). These perpendiculars divide the entire region into $n$ strips, each of width $\frac{\pi}{n}$. If we take each of these strips to be a rectangle of width $\frac{1}{n}\pi$ and altitude (counting the strips from the left) $\sin \frac{k\pi}{n}$, we see that the approximate area of the $k$th strip is

$$F_k \approx \frac{\pi}{n} \sin \frac{k\pi}{n}.$$

Thus, the area of the shaded region is approximately equal to

$$F \approx \frac{\pi}{n} \sum_{k=1}^{n} \sin \frac{k\pi}{n}.$$

This expression, by virtue of formula (28) in section 29, can be rewritten in the form

$$F \approx \frac{\pi}{n} \cdot \frac{\sin \frac{\pi}{2} \cdot \sin \frac{(n+1)\pi}{2n}}{\sin \frac{\pi}{2n}},$$

or (since $\sin \frac{\pi}{2} = 1$) in the form

$$F \approx \frac{\pi}{n} \cdot \frac{\sin \frac{(n+1)\pi}{2n}}{\sin \frac{\pi}{2n}}. \tag{33}$$

The exact expression for the area is the limit of the right-hand expression of (33) for an infinitely increasing $n$. This limit is found by means of the following considerations. Evidently,

$$\frac{(n + 1)\pi}{2n} = \frac{\pi}{2} + \frac{\pi}{2n},$$

so that the angle $\dfrac{(n + 1)\pi}{2n}$ tends to $\dfrac{\pi}{2}$; therefore, the sine of this angle must tend to 1:

$$\lim \left[ \sin \frac{(n + 1)\pi}{2n} \right] = 1. \tag{34}$$

On the other hand, the angle $\alpha_n = \dfrac{\pi}{2n}$ tends to zero and, thus, by virtue of formula (30) in section 31,

$$\lim \left[ \frac{\pi}{n} \cdot \frac{1}{\sin \dfrac{\pi}{2n}} \right] = \lim \left[ 2 \frac{\alpha_n}{\sin \alpha_n} \right] = 2. \tag{35}$$

From (34) and (35) (on the basis of the theorem concerning the limit of a product), we finally obtain

$$F = 2.$$

Hence, *the area of the region bounded by the sinusoid of half wave length and chord subtended by this half wave length is equal to 2.*

## 33. THE VOLUME ENCLOSED BY A ROTATING SINUSOID

Let us suppose that the sinusoid sketched in Fig. 24 is revolved about the $x$-axis. Let us find the volume $V$ of the solid bounded by the surface generated by rotating the portion of the sinusoid between $x = 0$ and $x = \pi$.

To do this, let us draw through the point $x = \dfrac{\pi}{2}$ a plane perpendicular to the $x$-axis. Clearly, this plane (Fig. 25) cuts the solid into two equal parts. Let us find the volume $V^*$ of the left half of the solid. To do this, we first divide the section of the $x$-axis from $x = 0$ to

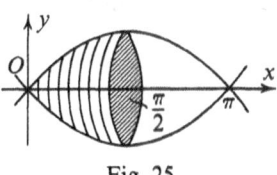

Fig. 25

$x = \frac{\pi}{2}$ into $n$ parts by means of the points

$$x_k = k\frac{\pi}{2n} \qquad (k = 1, 2, \ldots, n),$$

and through these points draw planes perpendicular to the $x$-axis. These planes intersect the surface of the solid in circles of radius (in the case of the $k$th plane)

$$r_k = \sin\frac{k\pi}{2n}.$$

If we treat the layer lying between the $(k - 1)$th and $k$th planes as a cylinder of radius $r_k$ and altitude $h = \frac{\pi}{2n}$, we find its volume to be

$$V_k \approx \pi r_k{}^2 h = \frac{\pi^2}{2n}\sin^2\frac{k\pi}{2n},$$

from which it follows that the total volume of the left-hand half of the solid is approximately equal to

$$V^* \approx \frac{\pi^2}{2n}\sum_{k=1}^{n}\sin^2\frac{k\pi}{2n}.$$

The exact expression for the volume is, therefore, the limit of the latter expression for an infinitely increasing $n$:

$$V^* = \lim\left[\frac{\pi^2}{2n}\sum_{k=1}^{n}\sin^2\frac{k\pi}{2n}\right. \tag{36}$$

To find this limit we shall employ an artificial method which permits us to shorten the calculation considerably. (It is because we had this method in mind that we began by considering, not the whole solid, but only its left half.) This method consists of the fol-

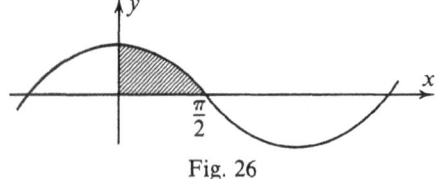

Fig. 26

lowing: Let us consider, just as we have for the sinusoid (32), the curve which is the graph of the function

$$y = \cos x. \tag{37}$$

Taking into account that

$$\cos x = \sin\left(x + \frac{\pi}{2}\right),$$

we see that the curve (37) is the same sinusoid (32), but displaced *to the left* along the $x$-axis by $\frac{\pi}{2}$ (Fig. 26).

Suppose now that we rotate *this* curve about the $x$-axis. It is clear that the volume of the solid formed by rotating the shaded region in Fig. 26 is equal to the volume $V^*$ of the left half of our original solid (for it is exactly equal to the volume of the *right* half of the original solid). On the other hand, were we to begin to calculate this volume by the method of summation, we could obviously arrive at a limit analogous to (36), with the substitution, however, of cosines for all the sines; that is, we should find that

$$V^* = \lim\left[\frac{\pi^2}{2n}\sum_{k=1}^{n}\cos^2\frac{k\pi}{2n}\right]. \tag{38}$$

Hence, $V^*$ can be represented in either one of two forms:

$$V^* = \lim\left[\frac{\pi^2}{2n}\sum_{k=1}^{n}\sin^2\frac{k\pi}{2n}\right] = \lim\left[\frac{\pi^2}{2n}\sum_{k=1}^{n}\cos^2\frac{k\pi}{2n}\right].$$

Adding these two expressions (which, clearly, it is possible to do, under the limit sign), we get

$$2V^* = \lim\left[\frac{\pi^2}{2n}\sum_{k=1}^{n}\left(\sin^2\frac{k\pi}{2n} + \cos^2\frac{k\pi}{2n}\right)\right]. \tag{39}$$

But

$$\sin^2\alpha + \cos^2\alpha = 1,$$

so that each term of the sum (39) is equal to 1, and, since the number of terms is $n$, we get

$$2V^* = \lim\left[\frac{\pi^2}{2n}\cdot n\right] = \lim\frac{\pi^2}{2} = \frac{\pi^2}{2},$$

since the limit of a constant is itself.

The volume $V^*$ of the left-hand half of the solid is thus found multiplied by two. But from the very beginning we have been interested in the volume of the whole solid, so we need not divide. Thus, the final result is

$$V = 2V^* = \frac{\pi^2}{2}. \tag{40}$$

Hence, *the volume of the solid bounded by the surface formed by rotating about its chord the portion of a sinusoid lying between $x = 0$ and $x = \pi$ is equal to $\frac{\pi^2}{2}$.*

## 34. MEAN VALUES

Let a certain quantity $y$ take on a finite number of values:

$$y_1, \quad y_2, \quad y_3, \quad \ldots, \quad y_n.$$

The arithmetic mean

$$y_* = \frac{y_1 + y_2 + \cdots + y_n}{n}$$

of these numbers $y_k$ is called the *mean value* of the quantity $y$. The usefulness of this quantity is to be found in two of its properties.

A. *If all the values of a quantity $y$ lie between numbers $m$ and $M$ then the mean value also lies between these two numbers; that is, if*

$$m \leq y_k \leq M \qquad (k = 1, 2, \ldots, n), \tag{41}$$

*then*

$$m \leq y_* \leq M.$$

B. *If all the values of a quantity $y$ are equal to the same number $h$, then the mean value is equal to this same number.*

Property B is obvious. To prove property A, it is necessary to sum all the inequalities obtained from (41); this gives

$$nm \leq \sum_{k=1}^{n} y_k \leq nM.$$

Dividing each term of this inequality by $n$, we obtain the desired result.

Along with the mean value $y_*$ of the quantity $y$, it is often useful to consider the *root mean square* $y^*$ of this same quantity. The root mean square is defined by the equality

$$y^* = \sqrt{\frac{y_1^2 + y_2^2 + \cdots + y_n^2}{n}}. \tag{42}$$

In other words, *the root mean square of the quantity $y$ is the square root of the mean value of the quantity $y^2$.*

It is easy to show that if all the values of the quantity $y$ are non-negative, its root mean square has the same properties $A$ and $B$ as the arithmetic mean. In fact, if

$$0 \leq m \leq y_k \leq M \qquad (k = 1, 2, \ldots, n),$$

then

$$m^2 \leq y_k^2 \leq M^2 \qquad (k = 1, 2, \ldots, n).$$

Summing these inequalities, dividing the result by $n$, and taking the square root, we obtain

$$m \leq y^* \leq M;$$

that is, we have shown that $y^*$ exhibits property A. Property B is obvious.

In the cases which we have considered, the quantity $y$ assumed a finite number of values. In practical problems one has, for the most part, to consider quantities which change continuously. In order to compute the means of such quantities, it is necessary to make use of the method of summation of infinitely small quantities. We shall illustrate this by an example from the field of physics.

## 35. EFFECTIVE CURRENT

Let us consider an alternating sinusoidal current

$$I = A \sin t, \tag{43}$$

where $t$ represents a number of units of time and $I$ represents a number of units of current. At different instants of time, the quantity $I$ has different values; let us assume the largest of these values to be $A$:

$$I_{max} = A. \tag{44}$$

An important role is played in electronics by the root mean square $I_e$ of the current during an oscillation, that is, from the time $t = 0$ to $t = 2\pi$. It can be shown that if the current is measured by an ammeter, this instrument will, in fact, register the quantity $I_e$. The quantity $I_e$ is called the *effective current*.

Let us compute $I_e$ for the current given by (43). We divide the interval of time from the instant $t = 0$ to the instant $t = 2\pi$ into $n$ small intervals of time; then

$$t_k = \frac{2\pi}{n} k \qquad\qquad (k = 1, 2, \ldots, n).$$

If the number $n$ is very large, we can assume, as an approximation, that for the interval of time from $t_{k-1}$ to $t_k$ the current does not change, and is equal to its value at the instant $t_k$:

$$I_k \approx A \sin \frac{2\pi}{n} k.$$

In other words, we assume that the current for a small moment of time is constant. Using this simplification, the effective current will be given by

$$I_e \approx \sqrt{\frac{\sum\limits_{k=1}^{n} I_k{}^2}{n}} = \sqrt{\frac{A^2 \sum\limits_{k=1}^{n} \sin^2 \frac{2\pi}{n} k}{n}}. \tag{45}$$

The true value of $I_e$ is the *limit* of the right-hand expression of (45) for an infinitely increasing $n$:

$$I_e = \lim \left[ A \sqrt{\frac{1}{n} \sum_{k=1}^{n} \sin^2 \frac{2\pi}{n} k} \right].$$

Let us find the limit of the expression under the radical:

$$\lim \left[ \frac{1}{n} \sum_{k=1}^{n} \sin^2 \frac{2\pi}{n} k \right]. \tag{46}$$

This can be done without any calculation in the following way.

Suppose that we are trying to find the volume of a solid formed by rotating one *wave length* of the sinusoid (32) about the $x$-axis.

If we employ the method of summation of infinitely small quantities, then, by repeating the reasoning of section 33, we can express this volume in the form

$$\lim \left[ \frac{2\pi^2}{n} \sum_{k=1}^{n} \sin^2 \frac{2k\pi}{n} \right].$$

On the other hand, this volume is clearly double the volume (40) of the solid found by rotating a *half wave length* of the sinusoid; that is, the volume we are seeking is equal to $\pi^2$. Hence,

$$\lim \left[ \frac{2\pi^2}{n} \sum_{k=1}^{n} \sin^2 \frac{2k\pi}{n} \right] = \pi^2. \tag{47}$$

It is easy to see that limit (46) is found from limit (47) by dividing through by $2\pi^2$, from which it follows that

$$\lim \left[ \frac{1}{n} \sum_{k=1}^{n} \sin^2 \frac{2k\pi}{n} \right] = \frac{1}{2}.$$

Hence,[1]

$$I_e = A \sqrt{\frac{1}{2}}. \tag{48}$$

This completes the solution of the problem.

Comparing formulas (48) and (44), we see that

$$I_{max} = I_e \sqrt{2};$$

that is, the maximum current is roughly one and a half times as great as the current registered by the ammeter.

---

[1] We use here the following theorem: If a variable quantity $x_n \geq 0$ tends to a limit $a$, then $\sqrt{x_n}$ tends to $\sqrt{a}$.

# Problems for Practice

We append a number of problems to afford independent practice in the methods explained. We urgently advise the reader to work out these problems. As Newton said: "In mathematics, problems are more useful than precepts."

1. Evaluate the sum

$$S_4 = \sum_{k=1}^{n} k^4.$$

*Answer.* $S_4 = \dfrac{n(n + 1)(2n + 1)(3n^2 + 3n - 1)}{30}.$

2. Determine the area of a right triangle by the summation method.

3. Find the area bounded by the x-axis, the curve $y = x^3$, and the straight line $x = 1$. *Answer.* $\dfrac{1}{4}$.

4. Find the limit $\lim \left[ \dfrac{1}{n^2} \sum\limits_{k=1}^{n} \sqrt{n^2 - k^2} \right]$ for an infinitely increasing $n$.

*Hint.* Determine the area of a quadrant of a circle by the summation of rectangular strips. *Answer.* $\dfrac{\pi}{4}$.

5. Proceeding from the result of the previous problem, find the volume of a cylinder by dividing it into rectangular slabs as shown in Fig. 9.

6. Determine the pressure of water on the wall of a cylindrical glass. *Answer.* $P = \pi R H^2$.

7. Find the work needed to pump water out of a full conical container the base of which is horizontal and *below* the vertex. *Answer.* $T = \dfrac{1}{4} \pi R^2 H^2$.

8. Determine the volume of an ellipsoid of revolution by direct calculation without reference to the principle of Cavalieri.

9. Find the volume of the ellipsoid formed by rotating the ellipse $\frac{x^2}{a^2} + \frac{y^2}{b^2} = 1$ about the $y$-axis. *Answer.* $V = \frac{4}{3}\pi a^2 b$.

10. What work is necessary in order to pump water from a hemispherical container with its plane surface downwards? *Answer.* $T = \frac{5}{12}\pi R^4$.

11. Determine the water pressure on the walls of a prismatic container the altitude of which is $H$ and the perimeter of whose base is $p$. *Answer.* $P = \frac{1}{2}pH^2$.

12. Using the result of Problem 1, find the volume of the solid bounded by the surface generated by rotating the parabola $y = ax^2$ about the $x$-axis and the plane perpendicular to the $x$-axis at distance $h$ from the origin. *Answer.* $V = \frac{1}{5}\pi a^2 h^5$.

13. Find the limit

$$\lim \left[ \frac{1}{n^{p+1}} \sum_{k=1}^{n} k^p \right]$$

for an infinitely increasing $n$ ($p$ is a positive whole number). *Answer.* $\frac{1}{p+1}$.

14. Find the limit

$$\lim \left[ \frac{1}{n} \sum_{k=1}^{n} \sqrt{\frac{k}{n}} \right]$$

for an infinitely increasing $n$.

*Hint.* Find the area of the curvilinear triangle $OQM$ (Fig. 19) by dividing it into strips parallel to the $x$-axis. *Answer.* $\frac{2}{3}$.

15. Find the sum

$$S = \sum_{k=1}^{n} \cos k\alpha.$$

$$\textit{Answer. } S = \frac{\sin \frac{2n+1}{2}\alpha}{2\sin \frac{\alpha}{2}} - \frac{1}{2}.$$

16. Using the result of Problem 15, find the area $F$ of the region bounded by the curve $y = \cos x$ and the $x$-axis, $0 \le x \le \frac{\pi}{2}$. *Answer.* $F = 1$.

A CATALOG OF SELECTED
# DOVER BOOKS
## IN SCIENCE AND MATHEMATICS

# Mathematics–Bestsellers

HANDBOOK OF MATHEMATICAL FUNCTIONS: with Formulas, Graphs, and Mathematical Tables, Edited by Milton Abramowitz and Irene A. Stegun. A classic resource for working with special functions, standard trig, and exponential logarithmic definitions and extensions, it features 29 sets of tables, some to as high as 20 places. 1046pp. 8 x 10 1/2.                                                0-486-61272-4

ABSTRACT AND CONCRETE CATEGORIES: The Joy of Cats, Jiri Adamek, Horst Herrlich, and George E. Strecker. This up-to-date introductory treatment employs category theory to explore the theory of structures. Its unique approach stresses concrete categories and presents a systematic view of factorization structures. Numerous examples. 1990 edition, updated 2004. 528pp. 6 1/8 x 9 1/4.                      0-486-46934-4

MATHEMATICS: Its Content, Methods and Meaning, A. D. Aleksandrov, A. N. Kolmogorov, and M. A. Lavrent'ev. Major survey offers comprehensive, coherent discussions of analytic geometry, algebra, differential equations, calculus of variations, functions of a complex variable, prime numbers, linear and non-Euclidean geometry, topology, functional analysis, more. 1963 edition. 1120pp. 5 3/8 x 8 1/2.       0-486-40916-3

INTRODUCTION TO VECTORS AND TENSORS: Second Edition--Two Volumes Bound as One, Ray M. Bowen and C.-C. Wang. Convenient single-volume compilation of two texts offers both introduction and in-depth survey. Geared toward engineering and science students rather than mathematicians, it focuses on physics and engineering applications. 1976 edition. 560pp. 6 1/2 x 9 1/4.                   0-486-46914-X

AN INTRODUCTION TO ORTHOGONAL POLYNOMIALS, Theodore S. Chihara. Concise introduction covers general elementary theory, including the representation theorem and distribution functions, continued fractions and chain sequences, the recurrence formula, special functions, and some specific systems. 1978 edition. 272pp. 5 3/8 x 8 1/2.
                                                                 0-486-47929-3

ADVANCED MATHEMATICS FOR ENGINEERS AND SCIENTISTS, Paul DuChateau. This primary text and supplemental reference focuses on linear algebra, calculus, and ordinary differential equations. Additional topics include partial differential equations and approximation methods. Includes solved problems. 1992 edition. 400pp. 7 1/2 x 9 1/4.                                            0-486-47930-7

PARTIAL DIFFERENTIAL EQUATIONS FOR SCIENTISTS AND ENGINEERS, Stanley J. Farlow. Practical text shows how to formulate and solve partial differential equations. Coverage of diffusion-type problems, hyperbolic-type problems, elliptic-type problems, numerical and approximate methods. Solution guide available upon request. 1982 edition. 414pp. 6 1/8 x 9 1/4.                          0-486-67620-X

VARIATIONAL PRINCIPLES AND FREE-BOUNDARY PROBLEMS, Avner Friedman. Advanced graduate-level text examines variational methods in partial differential equations and illustrates their applications to free-boundary problems. Features detailed statements of standard theory of elliptic and parabolic operators. 1982 edition. 720pp. 6 1/8 x 9 1/4.                                       0-486-47853-X

LINEAR ANALYSIS AND REPRESENTATION THEORY, Steven A. Gaal. Unified treatment covers topics from the theory of operators and operator algebras on Hilbert spaces; integration and representation theory for topological groups; and the theory of Lie algebras, Lie groups, and transform groups. 1973 edition. 704pp. 6 1/8 x 9 1/4.
                                                                 0-486-47851-3

A SURVEY OF INDUSTRIAL MATHEMATICS, Charles R. MacCluer. Students learn how to solve problems they'll encounter in their professional lives with this concise single-volume treatment. It employs MATLAB and other strategies to explore typical industrial problems. 2000 edition. 384pp. 5 3/8 x 8 1/2. 0-486-47702-9

NUMBER SYSTEMS AND THE FOUNDATIONS OF ANALYSIS, Elliott Mendelson. Geared toward undergraduate and beginning graduate students, this study explores natural numbers, integers, rational numbers, real numbers, and complex numbers. Numerous exercises and appendixes supplement the text. 1973 edition. 368pp. 5 3/8 x 8 1/2. 0-486-45792-3

A FIRST LOOK AT NUMERICAL FUNCTIONAL ANALYSIS, W. W. Sawyer. Text by renowned educator shows how problems in numerical analysis lead to concepts of functional analysis. Topics include Banach and Hilbert spaces, contraction mappings, convergence, differentiation and integration, and Euclidean space. 1978 edition. 208pp. 5 3/8 x 8 1/2. 0-486-47882-3

FRACTALS, CHAOS, POWER LAWS: Minutes from an Infinite Paradise, Manfred Schroeder. A fascinating exploration of the connections between chaos theory, physics, biology, and mathematics, this book abounds in award-winning computer graphics, optical illusions, and games that clarify memorable insights into self-similarity. 1992 edition. 448pp. 6 1/8 x 9 1/4. 0-486-47204-3

SET THEORY AND THE CONTINUUM PROBLEM, Raymond M. Smullyan and Melvin Fitting. A lucid, elegant, and complete survey of set theory, this three-part treatment explores axiomatic set theory, the consistency of the continuum hypothesis, and forcing and independence results. 1996 edition. 336pp. 6 x 9. 0-486-47484-4

DYNAMICAL SYSTEMS, Shlomo Sternberg. A pioneer in the field of dynamical systems discusses one-dimensional dynamics, differential equations, random walks, iterated function systems, symbolic dynamics, and Markov chains. Supplementary materials include PowerPoint slides and MATLAB exercises. 2010 edition. 272pp. 6 1/8 x 9 1/4. 0-486-47705-3

ORDINARY DIFFERENTIAL EQUATIONS, Morris Tenenbaum and Harry Pollard. Skillfully organized introductory text examines origin of differential equations, then defines basic terms and outlines general solution of a differential equation. Explores integrating factors; dilution and accretion problems; Laplace Transforms; Newton's Interpolation Formulas, more. 818pp. 5 3/8 x 8 1/2. 0-486-64940-7

MATROID THEORY, D. J. A. Welsh. Text by a noted expert describes standard examples and investigation results, using elementary proofs to develop basic matroid properties before advancing to a more sophisticated treatment. Includes numerous exercises. 1976 edition. 448pp. 5 3/8 x 8 1/2. 0-486-47439-9

THE CONCEPT OF A RIEMANN SURFACE, Hermann Weyl. This classic on the general history of functions combines function theory and geometry, forming the basis of the modern approach to analysis, geometry, and topology. 1955 edition. 208pp. 5 3/8 x 8 1/2. 0-486-47004-0

THE LAPLACE TRANSFORM, David Vernon Widder. This volume focuses on the Laplace and Stieltjes transforms, offering a highly theoretical treatment. Topics include fundamental formulas, the moment problem, monotonic functions, and Tauberian theorems. 1941 edition. 416pp. 5 3/8 x 8 1/2. 0-486-47755-X

**Browse over 9,000 books at www.doverpublications.com**

# Mathematics–Logic and Problem Solving

PERPLEXING PUZZLES AND TANTALIZING TEASERS, Martin Gardner. Ninety-three riddles, mazes, illusions, tricky questions, word and picture puzzles, and other challenges offer hours of entertainment for youngsters. Filled with rib-tickling drawings. Solutions. 224pp. 5 3/8 x 8 1/2.                    0-486-25637-5

MY BEST MATHEMATICAL AND LOGIC PUZZLES, Martin Gardner. The noted expert selects 70 of his favorite "short" puzzles. Includes The Returning Explorer, The Mutilated Chessboard, Scrambled Box Tops, and dozens more. Complete solutions included. 96pp. 5 3/8 x 8 1/2.                    0-486-28152-3

THE LADY OR THE TIGER?: and Other Logic Puzzles, Raymond M. Smullyan. Created by a renowned puzzle master, these whimsically themed challenges involve paradoxes about probability, time, and change; metapuzzles; and self-referentiality. Nineteen chapters advance in difficulty from relatively simple to highly complex. 1982 edition. 240pp. 5 3/8 x 8 1/2.                    0-486-47027-X

SATAN, CANTOR AND INFINITY: Mind-Boggling Puzzles, Raymond M. Smullyan. A renowned mathematician tells stories of knights and knaves in an entertaining look at the logical precepts behind infinity, probability, time, and change. Requires a strong background in mathematics. Complete solutions. 288pp. 5 3/8 x 8 1/2.

0-486-47036-9

THE RED BOOK OF MATHEMATICAL PROBLEMS, Kenneth S. Williams and Kenneth Hardy. Handy compilation of 100 practice problems, hints and solutions indispensable for students preparing for the William Lowell Putnam and other mathematical competitions. Preface to the First Edition. Sources. 1988 edition. 192pp. 5 3/8 x 8 1/2.                    0-486-69415-1

KING ARTHUR IN SEARCH OF HIS DOG AND OTHER CURIOUS PUZZLES, Raymond M. Smullyan. This fanciful, original collection for readers of all ages features arithmetic puzzles, logic problems related to crime detection, and logic and arithmetic puzzles involving King Arthur and his Dogs of the Round Table. 160pp. 5 3/8 x 8 1/2.

0-486-47435-6

UNDECIDABLE THEORIES: Studies in Logic and the Foundation of Mathematics, Alfred Tarski in collaboration with Andrzej Mostowski and Raphael M. Robinson. This well-known book by the famed logician consists of three treatises: "A General Method in Proofs of Undecidability," "Undecidability and Essential Undecidability in Mathematics," and "Undecidability of the Elementary Theory of Groups." 1953 edition. 112pp. 5 3/8 x 8 1/2.                    0-486-47703-7

LOGIC FOR MATHEMATICIANS, J. Barkley Rosser. Examination of essential topics and theorems assumes no background in logic. "Undoubtedly a major addition to the literature of mathematical logic." – *Bulletin of the American Mathematical Society.* 1978 edition. 592pp. 6 1/8 x 9 1/4.                    0-486-46898-4

INTRODUCTION TO PROOF IN ABSTRACT MATHEMATICS, Andrew Wohlgemuth. This undergraduate text teaches students what constitutes an acceptable proof, and it develops their ability to do proofs of routine problems as well as those requiring creative insights. 1990 edition. 384pp. 6 1/2 x 9 1/4.        0-486-47854-8

FIRST COURSE IN MATHEMATICAL LOGIC, Patrick Suppes and Shirley Hill. Rigorous introduction is simple enough in presentation and context for wide range of students. Symbolizing sentences; logical inference; truth and validity; truth tables; terms, predicates, universal quantifiers; universal specification and laws of identity; more. 288pp. 5 3/8 x 8 1/2.                    0-486-42259-3

**Browse over 9,000 books at www.doverpublications.com**

# Mathematics–Algebra and Calculus

VECTOR CALCULUS, Peter Baxandall and Hans Liebeck. This introductory text offers a rigorous, comprehensive treatment. Classical theorems of vector calculus are amply illustrated with figures, worked examples, physical applications, and exercises with hints and answers. 1986 edition. 560pp. 5 3/8 x 8 1/2.                0-486-46620-5

ADVANCED CALCULUS: An Introduction to Classical Analysis, Louis Brand. A course in analysis that focuses on the functions of a real variable, this text introduces the basic concepts in their simplest setting and illustrates its teachings with numerous examples, theorems, and proofs. 1955 edition. 592pp. 5 3/8 x 8 1/2.    0-486-44548-8

ADVANCED CALCULUS, Avner Friedman. Intended for students who have already completed a one-year course in elementary calculus, this two-part treatment advances from functions of one variable to those of several variables. Solutions. 1971 edition. 432pp. 5 3/8 x 8 1/2.                0-486-45795-8

METHODS OF MATHEMATICS APPLIED TO CALCULUS, PROBABILITY, AND STATISTICS, Richard W. Hamming. This 4-part treatment begins with algebra and analytic geometry and proceeds to an exploration of the calculus of algebraic functions and transcendental functions and applications. 1985 edition. Includes 310 figures and 18 tables. 880pp. 6 1/2 x 9 1/4.                0-486-43945-3

BASIC ALGEBRA I: Second Edition, Nathan Jacobson. A classic text and standard reference for a generation, this volume covers all undergraduate algebra topics, including groups, rings, modules, Galois theory, polynomials, linear algebra, and associative algebra. 1985 edition. 528pp. 6 1/8 x 9 1/4.                0-486-47189-6

BASIC ALGEBRA II: Second Edition, Nathan Jacobson. This classic text and standard reference comprises all subjects of a first-year graduate-level course, including in-depth coverage of groups and polynomials and extensive use of categories and functors. 1989 edition. 704pp. 6 1/8 x 9 1/4.                0-486-47187-X

CALCULUS: An Intuitive and Physical Approach (Second Edition), Morris Kline. Application-oriented introduction relates the subject as closely as possible to science with explorations of the derivative; differentiation and integration of the powers of x; theorems on differentiation, antidifferentiation; the chain rule; trigonometric functions; more. Examples. 1967 edition. 960pp. 6 1/2 x 9 1/4.                0-486-40453-6

ABSTRACT ALGEBRA AND SOLUTION BY RADICALS, John E. Maxfield and Margaret W. Maxfield. Accessible advanced undergraduate-level text starts with groups, rings, fields, and polynomials and advances to Galois theory, radicals and roots of unity, and solution by radicals. Numerous examples, illustrations, exercises, appendixes. 1971 edition. 224pp. 6 1/8 x 9 1/4.                0-486-47723-1

AN INTRODUCTION TO THE THEORY OF LINEAR SPACES, Georgi E. Shilov. Translated by Richard A. Silverman. Introductory treatment offers a clear exposition of algebra, geometry, and analysis as parts of an integrated whole rather than separate subjects. Numerous examples illustrate many different fields, and problems include hints or answers. 1961 edition. 320pp. 5 3/8 x 8 1/2.                0-486-63070-6

LINEAR ALGEBRA, Georgi E. Shilov. Covers determinants, linear spaces, systems of linear equations, linear functions of a vector argument, coordinate transformations, the canonical form of the matrix of a linear operator, bilinear and quadratic forms, and more. 387pp. 5 3/8 x 8 1/2.                0-486-63518-X

**Browse over 9,000 books at www.doverpublications.com**

# Mathematics–Probability and Statistics

BASIC PROBABILITY THEORY, Robert B. Ash. This text emphasizes the probabilistic way of thinking, rather than measure-theoretic concepts. Geared toward advanced undergraduates and graduate students, it features solutions to some of the problems. 1970 edition. 352pp. 5 3/8 x 8 1/2. 0-486-46628-0

PRINCIPLES OF STATISTICS, M. G. Bulmer. Concise description of classical statistics, from basic dice probabilities to modern regression analysis. Equal stress on theory and applications. Moderate difficulty; only basic calculus required. Includes problems with answers. 252pp. 5 5/8 x 8 1/4. 0-486-63760-3

OUTLINE OF BASIC STATISTICS: Dictionary and Formulas, John E. Freund and Frank J. Williams. Handy guide includes a 70-page outline of essential statistical formulas covering grouped and ungrouped data, finite populations, probability, and more, plus over 1,000 clear, concise definitions of statistical terms. 1966 edition. 208pp. 5 3/8 x 8 1/2. 0-486-47769-X

GOOD THINKING: The Foundations of Probability and Its Applications, Irving J. Good. This in-depth treatment of probability theory by a famous British statistician explores Keynesian principles and surveys such topics as Bayesian rationality, corroboration, hypothesis testing, and mathematical tools for induction and simplicity. 1983 edition. 352pp. 5 3/8 x 8 1/2. 0-486-47438-0

INTRODUCTION TO PROBABILITY THEORY WITH CONTEMPORARY APPLICATIONS, Lester L. Helms. Extensive discussions and clear examples, written in plain language, expose students to the rules and methods of probability. Exercises foster problem-solving skills, and all problems feature step-by-step solutions. 1997 edition. 368pp. 6 1/2 x 9 1/4. 0-486-47418-6

CHANCE, LUCK, AND STATISTICS, Horace C. Levinson. In simple, non-technical language, this volume explores the fundamentals governing chance and applies them to sports, government, and business. "Clear and lively ... remarkably accurate." – *Scientific Monthly*. 384pp. 5 3/8 x 8 1/2. 0-486-41997-5

FIFTY CHALLENGING PROBLEMS IN PROBABILITY WITH SOLUTIONS, Frederick Mosteller. Remarkable puzzlers, graded in difficulty, illustrate elementary and advanced aspects of probability. These problems were selected for originality, general interest, or because they demonstrate valuable techniques. Also includes detailed solutions. 88pp. 5 3/8 x 8 1/2. 0-486-65355-2

EXPERIMENTAL STATISTICS, Mary Gibbons Natrella. A handbook for those seeking engineering information and quantitative data for designing, developing, constructing, and testing equipment. Covers the planning of experiments, the analyzing of extreme-value data; and more. 1966 edition. Index. Includes 52 figures and 76 tables. 560pp. 8 3/8 x 11. 0-486-43937-2

STOCHASTIC MODELING: Analysis and Simulation, Barry L. Nelson. Coherent introduction to techniques also offers a guide to the mathematical, numerical, and simulation tools of systems analysis. Includes formulation of models, analysis, and interpretation of results. 1995 edition. 336pp. 6 1/8 x 9 1/4. 0-486-47770-3

INTRODUCTION TO BIOSTATISTICS: Second Edition, Robert R. Sokal and F. James Rohlf. Suitable for undergraduates with a minimal background in mathematics, this introduction ranges from descriptive statistics to fundamental distributions and the testing of hypotheses. Includes numerous worked-out problems and examples. 1987 edition. 384pp. 6 1/8 x 9 1/4. 0-486-46961-1

# Mathematics–Geometry and Topology

PROBLEMS AND SOLUTIONS IN EUCLIDEAN GEOMETRY, M. N. Aref and William Wernick. Based on classical principles, this book is intended for a second course in Euclidean geometry and can be used as a refresher. More than 200 problems include hints and solutions. 1968 edition. 272pp. 5 3/8 x 8 1/2. 0-486-47720-7

TOPOLOGY OF 3-MANIFOLDS AND RELATED TOPICS, Edited by M. K. Fort, Jr. With a New Introduction by Daniel Silver. Summaries and full reports from a 1961 conference discuss decompositions and subsets of 3-space; n-manifolds; knot theory; the Poincaré conjecture; and periodic maps and isotopies. Familiarity with algebraic topology required. 1962 edition. 272pp. 6 1/8 x 9 1/4. 0-486-47753-3

POINT SET TOPOLOGY, Steven A. Gaal. Suitable for a complete course in topology, this text also functions as a self-contained treatment for independent study. Additional enrichment materials make it equally valuable as a reference. 1964 edition. 336pp. 5 3/8 x 8 1/2. 0-486-47222-1

INVITATION TO GEOMETRY, Z. A. Melzak. Intended for students of many different backgrounds with only a modest knowledge of mathematics, this text features self-contained chapters that can be adapted to several types of geometry courses. 1983 edition. 240pp. 5 3/8 x 8 1/2. 0-486-46626-4

TOPOLOGY AND GEOMETRY FOR PHYSICISTS, Charles Nash and Siddhartha Sen. Written by physicists for physics students, this text assumes no detailed background in topology or geometry. Topics include differential forms, homotopy, homology, cohomology, fiber bundles, connection and covariant derivatives, and Morse theory. 1983 edition. 320pp. 5 3/8 x 8 1/2. 0-486-47852-1

BEYOND GEOMETRY: Classic Papers from Riemann to Einstein, Edited with an Introduction and Notes by Peter Pesic. This is the only English-language collection of these 8 accessible essays. They trace seminal ideas about the foundations of geometry that led to Einstein's general theory of relativity. 224pp. 6 1/8 x 9 1/4. 0-486-45350-2

GEOMETRY FROM EUCLID TO KNOTS, Saul Stahl. This text provides a historical perspective on plane geometry and covers non-neutral Euclidean geometry, circles and regular polygons, projective geometry, symmetries, inversions, informal topology, and more. Includes 1,000 practice problems. Solutions available. 2003 edition. 480pp. 6 1/8 x 9 1/4. 0-486-47459-3

TOPOLOGICAL VECTOR SPACES, DISTRIBUTIONS AND KERNELS, François Trèves. Extending beyond the boundaries of Hilbert and Banach space theory, this text focuses on key aspects of functional analysis, particularly in regard to solving partial differential equations. 1967 edition. 592pp. 5 3/8 x 8 1/2.
0-486-45352-9

INTRODUCTION TO PROJECTIVE GEOMETRY, C. R. Wylie, Jr. This introductory volume offers strong reinforcement for its teachings, with detailed examples and numerous theorems, proofs, and exercises, plus complete answers to all odd-numbered end-of-chapter problems. 1970 edition. 576pp. 6 1/8 x 9 1/4. 0-486-46895-X

FOUNDATIONS OF GEOMETRY, C. R. Wylie, Jr. Geared toward students preparing to teach high school mathematics, this text explores the principles of Euclidean and non-Euclidean geometry and covers both generalities and specifics of the axiomatic method. 1964 edition. 352pp. 6 x 9. 0-486-47214-0

**Browse over 9,000 books at www.doverpublications.com**